Lecture Notes in Physics

T0242180

The Lecture Notes in Physics

The series Lecture Notes in Physics (LNP), founded in 1969, reports new developments in physics research and teaching – quickly and informally, but with a high quality and the explicit aim to summarize and communicate current knowledge in an accessible way. Books published in this series are conceived as bridging material between advanced graduate textbooks and the forefront of research to serve the following purposes:

• to be a compact and modern up-to-date source of reference on a well-defined topic;

• to serve as an accessible introduction to the field to postgraduate students and nonspecialist researchers from related areas;

• to be a source of advanced teaching material for specialized seminars, courses and schools.

Both monographs and multi-author volumes will be considered for publication. Edited volumes should, however, consist of a very limited number of contributions only. Proceedings will not be considered for LNP.

Volumes published in LNP are disseminated both in print and in electronic formats, the electronic archive is available at springerlink.com. The series content is indexed, abstracted and referenced by many abstracting and information services, bibliographic networks, subscription agencies, library networks, and consortia.

Proposals should be sent to a member of the Editorial Board, or directly to the managing editor at Springer:

Dr. Christian Caron
Springer Heidelberg
Physics Editorial Department I
Tiergartenstrasse 17
69121 Heidelberg/Germany
christian.caron@springer-sbm.com

Antoine Llor

Statistical Hydrodynamic Models for Developed Mixing Instability Flows

Analytical "0D" Evaluation Criteria, and Comparison
of Single- and Two-Phase Flow Approaches

 Springer

Author

Antoine Llor
Commissariat à l'Énergie Atomique
PSRI/DPg/SESS
rue de la Fédération 31-33
75752 Paris Cedex 15, France
E-mail: antoine.llor@cea.fr

Antoine Llor, *Statistical Hydrodynamic Models for Developed Mixing Instability Flows*,
Lect. Notes Phys. 681 (Springer, Berlin Heidelberg 2005), DOI 10.1007/11531746

ISSN 0075-8450
ISBN-10 3-642-42180-6 Springer Berlin Heidelberg New York
ISBN-13 978-3-642-42180-8 Springer Berlin Heidelberg New York

Springer is a part of Springer Science+Business Media
springer.com
© Springer-Verlag Berlin Heidelberg 2005
Softcover re-print of the Hardcover 1st edition 2005

Typesetting: by the author using a Springer LaTeX macro package

Printed on acid-free paper   SPIN: 11531746   54/TechBooks   5 4 3 2 1 0

"The possible behavior of turbulence is much too complex and varied for a single parametrization to work in a broad range of situations. In our present state of understanding, these simple models will be based, in part, on good physics, in part, on bad physics, and in part, on shameless phenomenology. This is basically engineering."

J.L. Lumley,
in "Some comments on turbulence",
Phys. Fluids A **4**,203 (1992).

Preface

The availability of a turbulence model for gravitationally induced mixing instabilities that would be considered both "relevant" and "usable" by engineers has been and still is an important long-term goal at the DAM. The present work is just a first step in this endeavour, focused on defining a strategy and providing explicit meanings to the "relevance" and "usability" demands.

This work was started in mid 1997, at that time in the form of a few laboratory notes (Sects. 2.1, 3.1 and 7.2) for private use in an applications oriented study with a strong engineering emphasis. Originally, no new modeling developments were planned and a gravitationally driven mix–turbulence model was to be selected and adapted from the literature.

However, by mid 2000 a substantial amount of material (sometimes original or fundamental) had been gathered to support the development of a completely new model. The first version of the present report collected the notes accumulated during this time (hence his somewhat unusual format and content), and was published (in French) as an internal report in order to provide (hopefully) a convincing justification for starting this new model development. Moreover, it was aimed at becoming a basic and consistent training or reference text for non-specialist engineers in this field at the DAM (where the two-fluid flow "culture" is not very common).

In September 2001 an open version of the report was published (Report CEA-R-5983), and part of it was summarized in a communication to the 8th IWPCTM in December [54, 55]. Despite the time elapsed since its first publication and its many other defects, O. Schilling and some of his co-workers deemed it worth translating for use at Lawrence Livermore National Laboratory. The present version is adapted from this translation.

During the four years since the first publication, various elements have appeared in the literature and in the course of our own work (often under suggestions from other people) which cast new light on many parts of the present report: most have retained their relevance but some are now obsolete or even outright wrong. However, the report has not been updated except for part summaries, the new Sect. 9.3, and comments in footnotes; all these changes are labelled by Ⓤ (Update).

This work was long in the making. I therefore shall take first the opportunity to thank my supervisors, who expressed their support in various ways for over two years: F. Olive, E. Van Renterghem, and J. Duday. On a daily basis, I benefited from frequent, fertile, and crucial discussions on a number of points with, among others, P. Mignon, B. Desjardins, F. Dufour, X. Carlotti (dedicated reviewers of the manuscript), O. Simonin, A. Vallet, D.L. Youngs at AWE, B. Rebourcet, D. Bouche, J.-M. Ghidaglia, L. Chodorge, P.-H. Cournède, M. Legrand, J.-F. Haas, G. Pittion–Rossilon, G. Fiorese, C. Lion, and J.-M. Favre. The study was initially conducted using a numerical approach with DAM's k–ε model, for which J.-M. Pace and D. Souffland provided all the necessary help. I am also especially grateful to O. Schilling and his co-workers for their appreciation of this study, as well as all those at LLNL involved with the (painful) translation work.

Most of the analytic calculations presented here (lengthy and tedious though straightforward!), as well as most of the figures, were generated with the Mathematica® symbolic calculation software. Figures 4.1, 5.1, 6.4, and 8.1 were drawn using the IslandDraw® graphic software. The final document was composed using LaTeX.

Bruyères-le-Châtel   *Antoine Llor*
April 2005

Contents

**Part III Comparative Assessment of Models,
New Development Approaches**

1

Introduction

Many Dam applications involve phenomena of hydrodynamic mixing instabilities at interfaces between different materials. Simulation codes must include a suitable description of mixing dynamics in order to predict and reconstruct experiments [1]. For the most part, these instabilities rapidly evolve to a developed turbulent regime that can be described by a statistical hydrodynamic *model*. Choosing the correct model is *crucial* to the quality of a simulation.

The observed instabilities are combinations of three basic types: Kelvin–Helmholtz (due to shear), Richtmyer–Meshkov (due to a shock wave crossing an interface), and Rayleigh–Taylor (which occurs when the acceleration and the density gradient are in opposite directions). One of the first published statistical models that captured all these phenomena was a modification of the k–ε model [2], the standard form of which has become a classic in single-fluid hydro- and aerodynamic applications. As adapted to the specific needs of the Dam about ten years ago, this modified k–ε model, here called "Dam's k–ε" model [3–6], has been the subject of applications and publications too numerous to mention in the present report.

However, wide-ranging studies have revealed numerous inherent shortcomings of Dam's k–ε model, as evidenced by difficulties not only in reconstructing certain experiments, but even in maintaining consistency within a calculation (due to non-physical divergences of certain quantities in the codes) [7]. Attempts to improve the simulation of mixing instabilities thus have focused on refining the model or using other approaches, such as the k model [8–11], modified k–ε models [2,12–21], second-order models [13–17,22–28], multiscale models [26], two-fluid or even multifluid models [22–25,29–38], "bubble" models [39–48], phenomenological models [49–51], etc.

Because of this situation, studies for a new method of statistical modeling of mixing instabilities were undertaken at the Dam. However, starting from a theoretical formulation, the estimated cost of evaluating, adjusting, and validating a turbulent mixing model is too high to permit many "iterations" on its definition, even if the model was chosen "off-the-shelf" from the myriad of publications. The aim of the present report is to provide the *basic*

A. Llor: *Statistical Hydrodynamic Models for Developed Mixing Instability Flows*,
Lect. Notes Phys. **681**, 1 4 (2005)
www.springerlink.com

principles for establishing an effective strategy for choosing or developing a model. These principles were progressively made explicit in the course of a new model development [52,53].

The basic principles examined in the present report provide an answer to two broad but stringent modeling requirements:

- **A** – to specify in a just sufficient manner the set of *limiting cases* for the situations and physical quantities that a model must capture in order to be acceptable for the generally expected applications, and
- **B** – to provide qualitative, but discriminating, diagnostic tools that enable rapid evaluation of the usefulness and performance of a model for the limiting cases *at the earliest possible stage*, preferably during theoretical conceptualization.

Although these two points may seem general and thus trivial, they have a specific relevance for statistical models of mixing flows, and deserve being examined now in more detail.

Point A: in DAM applications, the turbulent mixing zones host many complex, coupled physical phenomena. Under these conditions, it is impossible to truly *validate* a hydrodynamic model. Of course, a very global verification of model performance may be achieved through some experiment reconstructions, but this is affected also by all other major parameters (equations of state, source terms, numerical schemes, etc.). It is thus necessary to isolate the influence of a model, using a set of *ideal* or canonical experiments specifically chosen to represent all the major hydrodynamic phenomena in an application. According to the hypotheses used here, some of which are subjective, a model can only be accepted if it simultaneously captures (with, of course, a *single* set of "universal" coefficients) the following aspects, which represent the "specification sheet":

1. The classic elementary phenomenology of developed incompressible single-fluid turbulence (relaxation, Reynolds stresses, turbulent diffusion, etc.).
2. The effects of shockless (isentropic) compressions.
3. The growth and energy balance of Kelvin–Helmholtz, Rayleigh–Taylor, and Richtmyer–Meshkov mixing zones in a developed self-similar regime between fluids of any types (as discussed in Sects. 4.2 and 4.3 in the limit of zero Atwood number).
4. The growth and energy balance of Rayleigh–Taylor mixing zones with self-similar acceleration as t^n in a developed regime for $-1 < n < \infty$ between fluids of any types (discussed in Sects. 5.3 and 6.2 in the limit of zero Atwood number).[1]

---

[1] Ⓤ Extension of these so called SSVARTs to $n < -2$ was found possible [54,55] and introduces new and important constraints on models, as discussed in Sect. 9.3.

5. The interaction of shock waves with turbulence and heterogeneous mixtures of fluids (discussed in Chap. 8).
6. The effects of mixing at various scales, especially at the atomic scale for miscible fluids (examined in Chap. 9).
7. The transition between laminar and turbulent regimes.

The first two items have been thoroughly investigated, primarily in aerodynamic applications, and there is no real need to revisit them here: k–ε models and their derivatives have proven dependable in these areas. Items 3 through 7, however, are more problematic, and none of the published models appears to fully satisfy these specifications. Only items 3 through 6 will be examined in this report, because item 7 pertains to the statistical nature of the models independently of their details, and has been the subject of numerous contributions (see, for example, [56] and the publications cited therein).

All these points have been raised and discussed separately in various publications, except item 4, which was added because of the specific needs of the DAM. In the applications to be studied, Rayleigh–Taylor instabilities can have any acceleration profile, even an erratic one, and it is virtually impossible to analyze a model under these conditions (even a "simple" model like the k–ε). It is therefore very useful to check a model behavior on a family of profiles that provides a function-decomposition basis and whose self-similarity almost always yields quick *analytical* results. However, it must be noted that there are yet no experimental data or numerical simulations for accelerations varying as t^n where $n \neq 0$.[2]

Point B: A qualitative method of model evaluation must be adapted to the desired refinement level of the phenomenology. Here, because the aim is to obtain a *global* estimate of growth and energy balance for the mixing zone, we shall consider a "0D" type of approach: for the various quantities of interest, only the *mean values over the mixing zone* will be considered, which makes it possible, by means of a few approximations, to simplify the partial differential equations to ordinary differential equations. The approximation procedure is greatly simplified in the limiting case of identical fluids (zero Atwood number), and the present study will consider this particular case only, described in Sect. 4.4, keeping in mind that an extension to asymmetric cases is possible.[3] The self-similar behavior of the flows under consideration leads to an additional simplification from ordinary differential equations to algebraic

---

[2] Ⓤ Preliminary results from numerical simulations of SSVARTs at $n = -1$ and 1 are now available [57].

[3] In previously published works, the implicit strategy is to verify that the model correctly reconstructs the RM and RT instabilities (usually for $n = 0$ only) over the entire range of Atwood numbers from 0 to 1. However, it was recently noted [38] that for reasons not entirely understood, this condition is not very restrictive, and most models show a suitable behavior with regard to variations of the Atwood number.

equations, which can be solved at least numerically and at best analytically (manually or using symbolic calculation programs).

Before proposing a new model, it was necessary to verify the relevance of the "0D" approach to validation. The two major classes of models in the literature are single-fluid and two-fluid. Each has important repercussions and is very different with regard to the structure and complexity of the numerical schemes and calculation codes. We·have thus illustrated here the potential of the "0D" approach by showing, along the specification criteria outlined above, its effectiveness in the assessment of two models representing these respective classes:

- DAM's k–ε model [5,6], and
- AWE's two-fluid model developed by D.L. Youngs at the AWE in the United Kingdom [33], which appears to provide one of the best compromises between simplicity and reliability, according to published results.

The theoretical framework for these models is reviewed in Chaps. 2 and 3, using notations that facilitate comparison, and their "0D" analysis is presented in Chaps. 5 and 6. Incidentally, the time saving provided by the "0D" approach is worth noting: using the reduction procedure described in Sect. 4.4 on models of equivalent complexity, the evaluation time for items 3 and 4 of the specification list can be reduced to a few engineer-weeks.

In Chap. 7, a critical comparison of DAM's k–ε and AWE's two-fluid models shows the reasons for favoring the two-fluid approach. This is followed by an evaluation of models of intermediate complexity, including "two-fluid" closures of the k–ε model, extended single-fluid models, and degenerate two-fluid models. Chapters 8 and 9 treat specification items 5 and 6, respectively. They provide general reflections, rather than evaluation tools like the "0D" approach, primarily because of the nature of the subjects, but also because of the lack of experimental, numerical or analytical reference data. They are, however, sufficient for defining and justifying modeling strategies, even if supplementary studies will be needed to make some closures explicit. On these points, the single- and two-fluid approaches appear essentially equivalent, with a slight advantage for the latter.

The conclusion summarizes all the findings and shows how they influence the ongoing development of a new two-fluid model at the DAM in collaboration with various other laboratories.

This supports the present choice of focusing on the behavior as a function of n, leaving open the effect of Atwood number for later investigation.

Part I

Short Primer on Basic Averaging
and Modeling Approaches
for Turbulent Mixing Instabilities

2

Single-Fluid Approach: Example of DAM's k–ε Model

2.1 Single-Fluid Statistical Equations

The statistical equations for the turbulent flow of an incompressible fluid are found in the literature [58], but they are adapted here to the specific needs of mixing flows, using more compact notations detailed in A.

We assume that the medium is a mixture of two components, indexed $+$ and $-$, that are dispersed on either microscopic scales (when the two fluids are miscible) or mesoscopic scales (i.e., intermediate between the macroscopic and microscopic scales associated with the overall flow and the particle mean free paths). In the hydrodynamic limit, the velocity at a given point, $\boldsymbol{u}$, is the velocity of the center of mass of the fluids at that point, while the interdiffusion flux, $\boldsymbol{\phi}^{\pm}$, takes into account the additional transport of the fluid components at the microscopic scale.

We have the mass conservation equations of the two species, and in the single-fluid approach, we also consider the momentum and total energy conservation equations. With the notations in A, these equations are written as:

$$
\left\|
\begin{array}{l}
\frac{\partial}{\partial t}\left(\ \varrho c^{\pm}\ \right) + \left(\ \varrho c^{\pm}\ \ u_j\right)_{,j} = -\left(\phi_j^{\pm}\right)_{,j} \\[4pt]
\frac{\partial}{\partial t}\left(\ \varrho u_i\ \right) + \left(\ \varrho u_i\ \ u_j\right)_{,j} = -\left(p\right)_{,i} - \left(\tau_{ij}\right)_{,j} \qquad\qquad + \varrho g_i \\[4pt]
\frac{\partial}{\partial t}\left(\ \varrho f\ \right) + \left(\ \varrho f\ \ u_j\right)_{,j} = -\left(pu_j\right)_{,j} - \left(\tau_{ij}u_i\right)_{,j} - \left(\theta_j\right)_{,j} + s + \varrho u_i g_i \\[4pt]
\frac{\partial}{\partial t}\left(\ \varrho e\ \right) + \left(\ \varrho e\ \ u_j\right)_{,j} = -p(u_j)_{,j} - \tau_{ij}(u_i)_{,j} - \left(\theta_j\right)_{,j} + \quad s \\[4pt]
\frac{\partial}{\partial t}\left(\varrho u_i u_i/2\right) + \left(\left(\varrho u_i u_i/2\right) u_j\right)_{,j} = -\left(p\right)_{,i} u_i - \left(\tau_{ij}\right)_{,j} u_i \qquad + \varrho u_i g_i
\end{array}
\right.
$$
$$ \tag{2.1} $$

where the balance of the internal and kinetic energy components is detailed in the last two equations. Since $\phi^+ + \phi^- = 0$, the sum of the mass conservation equations gives the total mass conservation:

$$
\frac{\partial}{\partial t}\varrho + \left(\varrho\, u_j\right)_{,j} = 0 \, . \tag{2.2}
$$

A. Llor: *Statistical Hydrodynamic Models for Developed Mixing Instability Flows*,
Lect. Notes Phys. **681**, 7 15 (2005)
www.springerlink.com

Most models and codes use this equation and one of the two equations in $c^{\pm}$, rather than the equivalent symmetric formulation in c^{+} and c^{-}. Hereinafter we shall retain the three equations in order to be consistent with the two-fluid approach presented in Chap. 3.

Because a full description of the flow cannot be obtained, neither analytically nor numerically, we shall settle for a more or less detailed statistical description of the system. The simplest description consists in deriving the *ensemble mean* of the equations, also known as the *Reynolds average*, with all the biases that are introduced therewith, in particular the non-equivalence with temporal or spatial means, the loss of information on fluctuations about means, the indirect comparison with the experiments, etc [59]. Denoting the Reynolds average with an overbar, we thus obtain:

$$
\left\|
\begin{array}{l}
\frac{\partial}{\partial t}\left(\ \overline{\varrho}\ \right) + \left(\ \overline{\varrho u_j}\ \right)_{,j} = 0 \\[4pt]
\frac{\partial}{\partial t}\left(\ \overline{\varrho c^{\pm}}\ \right) + \left(\ \overline{\varrho c^{\pm} u_j}\ \right)_{,j} = -(\overline{\phi_j^{\pm}})_{,j} \\[4pt]
\frac{\partial}{\partial t}\left(\ \overline{\varrho u_i}\ \right) + \left(\ \overline{\varrho u_i u_j}\ \right)_{,j} = -(\overline{p})_{,i} - (\overline{\tau_{ij}})_{,j} \qquad\qquad + \overline{\varrho} g_i \\[4pt]
\frac{\partial}{\partial t}\left(\ \overline{\varrho f}\ \right) + \left(\ \overline{\varrho f u_j}\ \right)_{,j} = -(\overline{p u_j})_{,j} - (\overline{\tau_{ij} u_i})_{,j} - (\overline{\theta_j})_{,j} + \overline{s} + \overline{\varrho u_i} g_i \\[4pt]
\frac{\partial}{\partial t}\left(\ \overline{\varrho e}\ \right) + \left(\ \overline{\varrho e u_j}\ \right)_{,j} = -\overline{p(u_j)_{,j}} - \overline{\tau_{ij}(u_i)_{,j}} - (\overline{\theta_j})_{,j} + \qquad \overline{s} \\[4pt]
\frac{\partial}{\partial t}\left(\overline{\varrho u_i u_i/2}\right) + \left(\overline{(\varrho u_i u_i/2)\, u_j}\right)_{,j} = -\overline{(p)_{,i}\, u_i} - \overline{(\tau_{ij})_{,j}\, u_i} \qquad + \overline{\varrho u_i} g_i
\end{array}
\right.
\tag{2.3}
$$

A key problem in the description of turbulence, and which is rarely discussed in classical textbooks [58], is that of the *effective fluid and the mean velocity*. Although the mean equations in (2.3) *are not hydrodynamic equations* like (2.1), they can be *interpreted*, at the price of some accommodations, as equations of an "effective fluid" having a "mean velocity." This velocity is defined in the general case as the *mean advection of the material*:

$$
\overline{\varrho u_j} = \overline{\varrho}\, U_j\ ,
\tag{2.4}
$$

$$
u_j'' = u_j - U_j\ .
\tag{2.5}
$$

where U is the so-called "Favre" mean of u, and u'' is the Favre fluctuation [58]. It can be shown [58] that *there is no mean more suited* to define a characteristic velocity of a "mean fluid.[1]" For interpreting experiments, however, corrections must be made to this velocity, which is not rigorously measured. Besides, the Favre average definition is such that a mean quantity contains the correlations between its fluctuations and the density fluctuations.

---

[1] ⓤ More specifically, U captures *all* the mass averaged transport, whatever its cause or nature (for instance, random or not). Alternatively, one could choose U to capture all the averaged transport of *any* other predefined quantity, for instance u_x, e, etc. This would be at the expense of an additionnal turbulent flux in the mass conservation equation but also of inconsistencies in the constitutive laws of the effective fluid (per mass momentum would differ from the retained average velocity, thermodynamic invariance could be lost, etc.).

As will be elaborated later, this aspect is fundamental in the case of flows with mixing between fluids of different densities.

The equations of mean motion are then rewritten using U in all advection terms and adding corrective terms. We thus obtain:

$$
\begin{aligned}
&\frac{\partial}{\partial t}(\,\bar{\varrho}\,) + (\,\bar{\varrho}\,U_j)_{,j} = 0 \\
&\frac{\partial}{\partial t}(\overline{\varrho c^{\pm}}) + (\overline{\varrho c^{\pm}}\,U_j)_{,j} = -(\overline{\phi_j^{\pm}})_{,j} \hspace{4cm} -\left(\overline{\varrho c^{\pm} u_j''}\right)_{,j} \\
&\frac{\partial}{\partial t}(\overline{\varrho}\,U_i) + (\overline{\varrho}\,U_i\,U_j)_{,j} = -(\overline{p})_{,i} - (\overline{\tau_{ij}})_{,j} \hspace{1.5cm} + \bar{\varrho}g_i - \left(\overline{\varrho u_i u_j''}\right)_{,j} \\
&\frac{\partial}{\partial t}(\,\overline{\varrho f}\,) + (\,\overline{\varrho f}\,U_j)_{,j} = -(\overline{pu_j})_{,j} - (\overline{\tau_{ij}u_i})_{,j} - (\overline{\theta_j})_{,j} + \bar{s} + \bar{\varrho}U_i g_i - \left(\overline{\varrho f u_j''}\right)_{,j}
\end{aligned}
$$

$$(2.6)$$

Introducing the Favre fluctuations of the concentration and of the total energy:

$$
\begin{aligned}
c^{\pm\,\prime\prime} &= c^{\pm} - \overline{\varrho c^{\pm}}/\bar{\varrho}\,, \\
f^{\prime\prime} &= f - \overline{\varrho f}/\bar{\varrho} = e^{\prime\prime} + u^2\!/_2 - \overline{\varrho u^2}/_{2\bar{\varrho}}\,,
\end{aligned}
\tag{2.7}
$$

the fluxes due to turbulent advections become:

$$
\begin{array}{ll}
\overline{\varrho c^{\pm} u_j''} = \overline{\varrho c^{\pm\,\prime\prime} u_j''} & \text{mean turbulent concentration flux,} \\
\overline{\varrho u_i u_j''} = \overline{\varrho u_i'' u_j''} = \overline{r_{ij}} & \text{mean Reynolds stress tensor,} \\
\overline{\varrho f u_j''} = \overline{\varrho f'' u_j''} & \text{mean turbulent flux of total energy.}
\end{array}
\tag{2.8}
$$

The evolution equations of these correlations can be obtained fairly easily [58], but they will not be considered here, since this report is primarily concerned with first-order models.

The "mean fluid" is distinguished from a real fluid by the contribution of a turbulent kinetic energy $k = \overline{u''^2}/_2$ in addition to the mean kinetic and internal energies. Thus the *total* kinetic energy is expressed as a function of the mean velocity as:

$$
\overline{\varrho u^2}/_2 = \bar{\varrho}U^2\!/_2 + \overline{\varrho k}\,,
\tag{2.9a}
$$

$$
k = \overline{u''^2}/_2 = r_{ii}/_{2\varrho}\,,
\tag{2.9b}
$$

and the total energy fluctuation is written as:

$$
f^{\prime\prime} = e^{\prime\prime} + k^{\prime\prime} + \boldsymbol{u}^{\prime\prime}.\boldsymbol{U}\,.
\tag{2.10}
$$

The energy balance, as detailed by the complementary equations in (2.3), must therefore be rederived to complete (2.6). It is obtained by subtracting the mean kinetic energy equation deduced from the mean momentum equation in (2.6):

$$
\begin{bmatrix}
\text{Mean} & \text{Favre} \\
\text{quantities} & \text{advections} \\
\frac{\partial}{\partial t}(\quad \overline{\varrho e} \quad)+(\quad \overline{\varrho e} \quad U_j)_{,j} \\
\frac{\partial}{\partial t}(\quad \overline{\varrho k} \quad)+(\quad \overline{\varrho k} \quad U_j)_{,j} \\
\frac{\partial}{\partial t}(\overline{\varrho}U_i U_i/2)+((\overline{\varrho}U_i U_i/2)U_j)_{,j} \\
\frac{\partial}{\partial t}(\quad \overline{\varrho f} \quad)+(\quad \overline{\varrho f} \quad U_j)_{,j}
\end{bmatrix}
$$

$$
=
\begin{bmatrix}
\multicolumn{3}{c}{\text{Mean motion}} \\
\text{Sound and RT} & \text{Dissipations} & \text{Sources} \\
-\ \overline{p}\,(\overline{u_i})_{,i} & -\ \overline{\tau_{ij}}(U_i)_{,j} -(\overline{\theta_j})_{,j} + & \overline{s} \\
-\ (\overline{p})_{,i}\,\overline{u_i''} & \cdot & \cdot \\
-\ (\overline{p})_{,i}\,U_i & -\ (\overline{\tau_{ij}})_{,j}\,U_i & +\ \overline{\varrho}U_i g_i \\
-\ (\overline{p\,u_i})_{,i} & -\ (\overline{\tau_{ij}U_i})_{,j} -(\overline{\theta_j})_{,j} & +\,\overline{s}+\overline{\varrho}U_i g_i
\end{bmatrix}
$$

$$
+
\begin{bmatrix}
\multicolumn{5}{c}{\text{Turbulent motion}} \\
\text{Noise} & \text{Pressure} & \text{Dissipation} & \text{Turbulent fluxes} & \text{Production} \\
-\overline{p'(u_i'')}_{,i} & \cdot & -\ \overline{\tau_{ij}(u_i'')}_{,j} & -\ (\overline{\varrho e''u_j''})_{,j} & \cdot \\
+\overline{p'(u_i'')}_{,i} & -(\overline{p'u_i''})_{,i} & -(\overline{\tau_{ij}})_{,j}\,u_i'' & -(\overline{\varrho k''u_j''})_{,j} & -\ \overline{\tau_{ij}}(U_i)_{,j} \\
\cdot & \cdot & \cdot & & -\ (\overline{\tau_{ij}})_{,j}\,U_i \\
\cdot & -(\overline{p'u_i''})_{,i} & -(\overline{\tau_{ij}u_i''})_{,j} & -(\overline{\varrho f''u_j''})_{,j} &
\end{bmatrix}
\qquad (2.11)
$$

Partial versions of this balance are given in [58] and [60]. The pressure was decomposed into mean and fluctuating components because the mean momentum equation in (2.6) contains only the mean pressure, while the pressure fluctuation terms are associated with various phenomena which require specific modeling approaches as shown in B. The names of the various terms in (2.11) are readily descriptive, except for "Sound and RT," which will be discussed in Sect. 7.2.

All the statistical equations presented so far have not been subjected to any approximation and are valid regardless of *the nature and properties of the described fluctuating fields*: velocity or other, values of the characteristic numbers (Reynolds number, in particular), etc. These equations alone cannot be used to calculate the motion. The complementary terms introduced by the Reynolds average must be provided by closures and models, through either algebraic or evolution equations, that in most cases are specifically adapted to a particular type of flow.

2.2 An Elementary Single-Fluid Closure: The Incompressible k–ε Model

It would be impossible here to discuss at length the various types of models of the complementary terms in (2.6) and (2.11). Such discussions may be found in general texts [61–63], or in [3–6,8–11,64–68] for flows and codes used by the DAM. However, the major steps of the simplest closures (mostly algebraic), which form the basis of the k–ε model used in the DAM codes [3–6,64,65], will be summarized here in order to show both the reasons for the difficulties that have been encountered in applications, and the potential advantages of the two-fluid approach.[2]

The k–ε model was first developed to describe an incompressible homogeneous fluid according to the 8 steps described below [62,69–76]:

1. The fluid being homogeneous and incompressible, the Favre average is identical to the Reynolds average: $(u_i'')_{,i} = 0$ and $\overline{u_i''} = 0$. Thus in (2.11) we can cancel the terms "Noise" and "Sound and RT" in the equations of the internal and turbulent kinetic energies.

2. We assume a flow with developed turbulence in spectral quasi-equilibrium [59,77]. In other words, the process of turbulent energy transfer by the nonlinearities of the Navier–Stokes equation, from the large scales of production by the mean flow to the small scales of dissipation where viscosity phenomena are no longer negligible, is in a quasi-steady state and extends over a range of scales that is sufficiently broad to create a Kolmogorov inertial cascade. This requires not only a high Reynolds number, but also the absence of transient phenomena such as a transition to turbulence, a singular point in the flow, etc. Therefore, we can disregard the terms in (2.11) labeled "Dissipations by mean motion," as well as the turbulent kinetic energy flux $\overline{\tau_{ij} u_i''}$ contained in the dissipation term.

3. We apply a closure, here called "Boussinesq–Reynolds,[3]" to the turbulent fluxes (2.8) [61, §4.7]. This simple approach is based on the analogy, for mesoscopic scales, with the first gradient closures of statistical thermodynamics:

$$\overline{\varrho a u_j''} = \bar{\varrho}\,\overline{a' u_j'} \overset{\text{m}}{=} -\bar{\varrho}\, D_t^a\,(\bar{a})_{,j}\,, \tag{2.12}$$

(we use the Reynolds fluctuations of the various quantities here because the fluid is incompressible and homogeneous). The diffusion coefficient D_t^a is then expressed in terms of the mean quantities of the flow (zero equation models) or using the kinetic turbulent energy (models with one or more equations) and is assumed to be scalar in simple models:

$$D_t^a = \frac{C_\mu}{\sigma_a}\lambda_i\sqrt{\bar{k}}\,. \tag{2.13}$$

---

[2] The models developed at the DAM as precursors to the current k–ε model are reviewed in [66,67].

[3] Depending on the source, this closure is also called "in first gradients," "Reynolds analogy," or "Boussinesq."

λ_i is the integral length scale of turbulence, which represents the mean free path due to turbulent advection, and the dimensionless constants σ_a are the turbulent Schmidt–Prandtl numbers associated with the various quantities a (knowing that $\sigma_u = 1$ from the definition of C_μ). λ_i has then to be given in terms of the mean values of the flow (one-equation models) or in terms of supplementary turbulent quantities as in the case of the k–ε model (models with two or more equations). The Boussinesq–Reynolds closure is only acceptable if the gradient scale of $\bar{a}$ is larger than the integral length scale, which requires the introduction of limiters [61, §4.6] or more complicated formulations (for example, nonlinear) if this so-called realizability condition is not met.

4. In the case of the Reynolds tensor, for which $a = u_i$, the Boussinesq–Reynolds closure (2.12) is applied to its deviator portion using the symmetrized strain-rate tensor:

$$\overline{u_i' u_j'} \stackrel{m}{=} 2/3\,\overline{k}\,\delta_{ij} - C_\mu\,\lambda_i\,\sqrt{\overline{k}}\,[(U_i)_{,j} + (U_j)_{,i}]\,. \tag{2.14}$$

5. The term of turbulent kinetic energy diffusion due to the pressure–velocity-fluctuation correlations, $\overline{p'u_i''} = \overline{pu_i'}$ in (2.11), is also modeled by a Boussinesq–Reynolds closure [61, §6.3]. Regrouping with the turbulent flux of $\overline{k}$, we obtain an effective coefficient σ_k, which encompasses the set of diffusive turbulent effects.

6. We introduce the characteristic quantity $\bar{\varepsilon}$ of the dissipation (or transfer) of large turbulent structures [59]. This makes it possible to define the integral length scale of turbulence as:

$$\lambda_i = \overline{k}^{3/2}/\bar{\varepsilon}\,. \tag{2.15}$$

This is possible because of the spectral quasi-equilibrium mentioned in item 2, by means of which the larger portion of the turbulent kinetic energy is concentrated at large scales. Furthermore, according to this assumption, we can identify $\bar{\varepsilon}$ with the dissipation into internal energy in (2.11) $-\overline{\tau_{ij}(u_i'')_{,j}}/\bar{\varrho}$.

7. An equation for the evolution of $\bar{\varepsilon}$ is constructed, partly by analysis of the statistical equation of $\overline{\tau_{ij}(u_i'')}_{,j}$ [69–76], but mainly by analogy with the modeled equation of $\overline{k}$ [61, §7], [77,78]. Indeed, dissipation is created and destroyed in parallel with turbulence, and each production and dissipation term in the equation of $\overline{k}$ is thus "mirrored" into a term in the equation of $\bar{\varepsilon}$ by applying a factor $C_\varepsilon \bar{\varepsilon}/\overline{k}$.

8. The various constants of the model (turbulent viscosity constant C_μ, Schmidt–Prandtl numbers σ, and mirror coefficients of the $\bar{\varepsilon}$ equation C_ε) are adjusted on canonical experiments (dissipation and shear of grid turbulence, turbulent boundary layer, etc.) or are constrained by consistency conditions (equality of the supports of $\overline{k}$ and $\bar{\varepsilon}$, etc.) [62]. These are compiled in Table 2.1 in Sect. 2.3.

Because the equations in $\bar{k}$ and $\bar{\varepsilon}$ are homogeneous, these quantities must be given non-zero initial values for at least one point so that the model can develop turbulence. The transition from laminar to turbulent in unsteady flows must be represented by a special procedure which is thus *external* to the k–ε model. Similarly, specific boundary conditions must be given at the surfaces of solid bodies, but this aspect will not be discussed here, because DAM applications usually involve free flows.

2.3 An Extension to Variable Densities: DAM's k–ε Model

Among the elementary models of incompressible turbulence, the k–ε model is popular in engineering applications because of its simplicity, robustness, and efficiency in situations of quasi-isotropy and spectral quasi-equilibrium. However, the extension of this model for compressible or heterogeneous flows remains controversial because it involves complementary terms having more complicated closures [2,12,60,63,79,80]. The version usually implemented in DAM codes is obtained by correcting and adding steps to the ones described above in Sect. 2.2 [3–6,64,65]:

9. Because of the particular behavior of turbulent acoustic modes, the "Noise" terms can be disregarded in DAM applications, as is discussed in B. Except for the diffusion term $\overline{p'u_i''}$ closed in paragraph 5 above, only the mean pressure $\bar{p}$ is retained in the set of equations.

10. The Boussinesq–Reynolds closure is extended to the compressible case as follows:[4]

$$\overline{\varrho a u_j''} = \overline{\varrho a'' u_j''} = \overline{\varrho a' u_j''} \overset{\mathrm{m}}{=} -\bar{\varrho}\, D_t^a\, (\tilde{a})_{,j}\;, \quad \text{where:}\;\; \tilde{a} = \overline{\varrho a}/\bar{\varrho}\,, \qquad (2.16)$$

and where $\bar{k}$ and $\bar{\varepsilon}$ are replaced by $\tilde{k}$ and $\tilde{\varepsilon}$ in the expressions for D_t^a and λ_i in (2.13) and (2.15). In the homogeneous incompressible case, (2.12) and (2.16) are identical.

11. The "Sound and RT" term in the equation for $\tilde{k}$ in (2.11), which is traditionally called the "Rayleigh–Taylor production" or "enthalpic production" term, and denoted here as π_{RT}, is closed by a Boussinesq–Reynolds type method (2.16) applied to the turbulent mass flux, $\overline{u_i''}$:

$$\pi_{RT} = -(\bar{p})_{,i}\overline{u_i''} = -(\bar{p})_{,i}\,\overline{\varrho v u_i''} \overset{\mathrm{m}}{=} -\frac{C_\mu}{\sigma_\varrho}\frac{\tilde{k}^2}{\tilde{\varepsilon}}\frac{(\bar{p})_{,i}(\bar{\varrho})_{,i}}{\bar{\varrho}}\;. \qquad (2.17)$$

Some authors [2–6,8–21] attribute to this term the production of turbulent kinetic energy during a Rayleigh–Taylor mixing instability, for which

---

[4] For the Reynolds tensor, only the traceless symmetrized portion is involved, as in (2.14).

$(\overline{p})_{,i}(\overline{\varrho})_{,i} < 0$. Because the destruction of $\tilde{k}$ is too strong in the case of demixing (with the possibility of *non-dissipative* destruction), π_{RT} is canceled if $(\overline{p})_{,i}(\overline{\varrho})_{,i} > 0$. Finally, a limiter is applied to the closure of the turbulent mass flux if the length of the density gradient is smaller than the integral length scale; this improvement is even more important for the "Rayleigh–Taylor term" than for turbulent fluxes, because it is an energy *production* term. The currently recommended limitation is:

$$\|\overline{u''}\| \leq \frac{2}{3}\sqrt{n_i \frac{\overline{\varrho u_i'' u_j''}}{\overline{\varrho}} n_j} \,, \tag{2.18}$$

where the Reynolds tensor closure has already been limited and $\boldsymbol{n}$ is the unit vector in the $\overline{\boldsymbol{u''}}$ direction.

12. The "Sound and RT" term in the equation of $\tilde{e}$ in (2.11) is decomposed by comparing it with the associated terms in the equations of mean kinetic energy and turbulent energy to obtain the term π_{RT}:

$$-\overline{p}(\overline{u_i})_{,i} = -\overline{p}(U_i)_{,i} - (\overline{p u_i''})_{,i} - \pi_{RT} \,. \tag{2.19}$$

Here π_{RT} is also modeled as in (2.17) to ensure the conservation of energy, while the flux $\overline{p u_i''}$ is simply disregarded. This latter approximation is justified for a first-gradient closure with adiabatic corrections (see E), but is not used in certain cases, where $\overline{u_i''}$ is then closed as in (2.17).

13. By extending the heuristic derivation method of the above-mentioned equation of $\overline{\varepsilon}$, a mirror term of "Rayleigh–Taylor production" is introduced into this equation with a coefficient of $C_{\varepsilon 0}\tilde{\varepsilon}/\tilde{k}$.

14. For a better reconstruction of the effects of rapid compressions (shocks), the mirror term of production by the Reynolds tensor is sometimes separated into pure shear and pure compression, associated with different coefficients $C_{\varepsilon 1}$ and $C_{\varepsilon 3}$.

'15. The physical and numerical treatment of shocks is realized by adding to the pressure an artificial viscosity stress (often isotropic) in all the equations (except the equation of state, of course). The artificial viscosity is calculated using classical Von Neumann–Richtmyer formulas [81] applied to the gradient of the Favre averaged velocity.

16. The supplementary coefficients of the model (σ_ϱ, $C_{\varepsilon 0}$, and possibly $C_{\varepsilon 3}$) are adjusted to capture shock tube experiments. The set of coefficients for the model is summarized in Table 2.1:

To summarize, the equations of DAM's k–ε model are [5,6]:

Table 2.1. Recommended constants for DAM's k–ε model [68]

Incompressible							Compressible		
C_μ	σ_c	σ_e	σ_k	σ_ε	$C_{\varepsilon 2}$	$C_{\varepsilon 1}$	σ_ϱ	$C_{\varepsilon 0}$	$C_{\varepsilon 3}$
.09	.7	.9	1.0	1.3	1.9	1.47	2.0	.85	.0

$$\begin{cases}
\frac{\partial}{\partial t}(\ \bar{\varrho}\ ) + (\ \bar{\varrho}\ U_j)_{,j} = 0 , \\[1.5ex]
\frac{\partial}{\partial t}(\bar{\varrho}\widetilde{c^{\pm}}) + (\bar{\varrho}\widetilde{c^{\pm}}U_j)_{,j} = -(\Phi_j^{c\pm})_{,j} , \\[1.5ex]
\frac{\partial}{\partial t}(\bar{\varrho}U_i) + (\bar{\varrho}U_i\,U_j)_{,j} = -(R_{ij})_{,j} - (\bar{p})_{,i} + \bar{\varrho}g_i , \\[1.5ex]
\frac{\partial}{\partial t}(\ \bar{\varrho}\tilde{e}\ ) + (\ \bar{\varrho}\tilde{e}\ U_j)_{,j} \overset{m}{=} -(\Phi_j^e)_{,j} - \bar{p}(U_i)_{,i} - \pi_{RT} + \bar{\varrho}\tilde{\varepsilon} + \bar{s} , \\[1.5ex]
\frac{\partial}{\partial t}(\ \bar{\varrho}\tilde{k}\ ) + (\ \bar{\varrho}\tilde{k}\ U_j)_{,j} \overset{m}{=} -(\Phi_j^k)_{,j} + \pi_{KH} + \pi_{RT} - \bar{\varrho}\tilde{\varepsilon} , \\[1.5ex]
\frac{\partial}{\partial t}(\ \bar{\varrho}\tilde{\varepsilon}\ ) + (\ \bar{\varrho}\tilde{\varepsilon}\ U_j)_{,j} \overset{m}{=} -(\Phi_j^\varepsilon)_{,j} + C_{\varepsilon 1}\frac{\tilde{\varepsilon}}{\tilde{k}}\pi_{KH} \\[1.5ex]
\qquad\qquad + C_{\varepsilon 0}\frac{\tilde{\varepsilon}}{\tilde{k}}\pi_{RT} - C_{\varepsilon 2}\bar{\varrho}\frac{\tilde{\varepsilon}^2}{\tilde{k}} - C_{\varepsilon 3}\tilde{\varepsilon}(U_i)_{,i} ,
\end{cases} \tag{2.20}$$

where the turbulent fluxes are closed by:

$$\Phi_j^a \overset{m}{=} -\bar{\varrho}\frac{C_\mu}{\sigma_a}\frac{\tilde{k}^2}{\tilde{\varepsilon}}(\tilde{a})_{,j} , \tag{2.21a}$$

$$R_{ij} \overset{m}{=} \bar{\varrho}\frac{2\tilde{k}}{3}\delta_{ij} - \bar{\varrho}C_\mu\frac{\tilde{k}^2}{\tilde{\varepsilon}}[(U_i)_{,j} + (U_j)_{,i} - {}^2\!/_3(U_k)_{,k}\delta_{ij}] , \tag{2.21b}$$

π_{RT} is given in (2.17), and:

$$\pi_{KH} = -R_{ij}(U_i)_{,j} . \tag{2.22}$$

As in the incompressible case above on p. 13, these equations are homogeneous and cannot spontaneously cause the transition to a turbulent flow starting from a laminar situation where $\tilde{k}$ and $\tilde{\varepsilon}$ are zero. For mixing flows, a mixing zone where $\tilde{k}$ and $\tilde{\varepsilon}$ differ from zero must be explicitly defined at the initial state of the calculation [56, 64, 65].

The incompressible k-ε model [69–76] has acquired some status as a universal model:[5] the limitations are known and the constants were established many years ago, with variations of the order of $\pm 15\%$ and $\pm 5\%$ for σ and C_ε, respectively, depending on the authors and the particular flows under consideration [61, 62]. The robustness and simplicity of this model have made it a preferred tool for fluid mechanics engineers. However, it is nowhere near that status in its compressible form, because numerous points give rise to physical inconsistencies: the values of the constants in the compressible extension, the extension of Boussinesq-Reynolds closures, the application to mixing flows, and the consideration of shocks. These aspects will be illustrated in Chap. 5 using as example the restitution of the Rayleigh–Taylor instability and will be analyzed more generally in Chap. 7.

---

[5] ⓒ Actually, the k-ε model is not exactly "universal" in the physical sense, and it is regarded as limited in the academic community. Yet, it does enjoy a very high status in engineering, well beyond applications to simple shear boundary layers, because it provides reasonable qualitative results for low cost and complexity, and above all, with excellent robustness. Furthermore, a vast majority of sophisticated models do embed some form of k-ε as a limiting case.

3

Two-Fluid Approach: Example of AWE's Model

3.1 Two-Fluid Statistical Equations

Two-fluid descriptions often use explicitly separate treatments for the conservation equations in each fluid, along with expressions for exchange terms that can become complicated [82–86]. We will re-examine the conservation equations (2.1) for a single global fluid with variable properties depending on the position: equation of state, constitutive laws, thermodynamic variables, etc. The various quantities will thus reflect variations due to both flow and composition of the fluid, and the properties of individual fluid components will be selected by taking the product with the local mass fractions c^+ and $c^- = 1 - c^+$. All possible mixing cases can be treated within this formalism (immiscible fluids, molecular interdiffusions, surface tensions, etc.) and, under various closure assumptions, lead to the usual models (such as reviewed in [83–86], for example).

We note that the mean mass fraction (Favre average) is:

$$C^\pm = \widetilde{c^\pm} = \overline{c^\pm \varrho}/\overline{\varrho} \, . \tag{3.1}$$

In addition, in the case where the two fluids are not miscible, the $c^\pm$ are *indicator functions* of the fluids. Thus we can express the mean volume fractions[1] and densities of the fluids as:

$$\alpha^\pm = \overline{c^\pm} \, , \tag{3.2a}$$

$$\varrho^\pm = \overline{c^\pm \varrho}/\overline{c^\pm} \, . \tag{3.2b}$$

From these we deduce the expression for the local Atwood number:

---

[1] According to (3.2b), the $\alpha^\pm$ should be more appropriately designated as "presence probabilities," but in the limit of homogeneous flows they coincide with mean volume fractions. Common usage in publications on two-fluid flow models is to retain the latter terminology, whatever the averaging process.

A. Llor: *Statistical Hydrodynamic Models for Developed Mixing Instability Flows*,
Lect. Notes Phys. **681**, 17 28 (2005)
www.springerlink.com

$$A = \frac{\varrho^+ - \varrho^-}{\varrho^+ + \varrho^-}\,. \tag{3.3}$$

Analogously, the mean value of any per mass quantity a (such as internal energy) for each fluid will be given by:

$$A^\pm = \overline{c^\pm \varrho a}\big/\overline{c^\pm \varrho}\,. \tag{3.4}$$

We extend all these definitions to the case of miscible fluids. Thus (3.3) correctly provides a zero Atwood number in the limit of a molecular mixture, for which all the correlations with $c^\pm$ disappear, $\overline{c^+a}/\overline{c^+} = \overline{c^-a}/\overline{c^-}$.[2]

Multiplying equations (2.1) by $c^\pm$, we obtain the non-averaged two-fluid equations:[3]

$$\frac{\partial}{\partial t}\left(c^\pm\ \varrho\ \right) + \left(c^\pm\ \varrho u_j\right)_{,j} = -(\phi_j^\pm)_{,j}$$

$$\frac{\partial}{\partial t}\left(c^\pm\ \varrho u_i\right) + \left(c^\pm\ \varrho u_i\right)_{,j} = -(c^\pm p)_{,i} - (c^\pm \tau_{ij})_{,j} + (c^\pm)_{,i} p + (c^\pm)_{,j} \tau_{ij} + c^\pm \varrho g_i - (\phi_j^\pm)_{,j} u_i$$

$$\frac{\partial}{\partial t}\left(c^\pm\ \varrho f\right) + \left(c^\pm\ \varrho f u_j\right)_{,j} = -(c^\pm p u_j)_{,j} - (c^\pm \tau_{ij} u_i)_{,j} - (c^\pm \theta_j)_{,j} + (c^\pm)_{,j} p u_j + (c^\pm)_{,j} \tau_{ij} u_i + (c^\pm)_{,j} \theta_j + c^\pm s + c^\pm u_i \varrho g_i - (\phi_j^\pm)_{,j} f$$

$$\frac{\partial}{\partial t}\left(c^\pm\ \varrho e\right) + \left(c^\pm\ \varrho e u_j\right)_{,j} = -c^\pm p(u_j)_{,j} - c^\pm \tau_{ij}(u_i)_{,j} - (c^\pm \theta_j)_{,j} + (c^\pm)_{,j} \theta_j + c^\pm s + (\phi_j^\pm)_{,j} e$$

$$\frac{\partial}{\partial t}\left(c^\pm \varrho \frac{u_i u_i}{2}\right) + \left(c^\pm \varrho \frac{u_i u_i}{2} u_j\right)_{,j} = -(c^\pm p)_{,i} u_i - (c^\pm \tau_{ij})_{,j} u_i + (c^\pm)_{,j} p u_j + (c^\pm)_{,j} \tau_{ij} u_i + c^\pm u_i \varrho g_i - (\phi_j^\pm)_{,j} \frac{u_i u_i}{2} \tag{3.5}$$

In the right hand sides, we have included terms in $(c^\pm)_{,j}$, which represent *exchanges between fluids*, because $(c^+)_{,j} + (c^-)_{,j} = 0$. For immiscible fluids, the mass fractions $c^\pm$ can take values 0 or 1 only, and thus their gradients are vectorial Dirac distribution sheets at the interfaces, which provide a description of the boundary conditions between the fluids.

As in Sect. 2.1, we take the Reynolds average and introduce the two mean velocities, $U^\pm$, which are defined by the mean advections of materials + and $-$:

$$\overline{c^\pm \varrho u_j} = \overline{c^\pm \varrho}\, U_j^\pm , \qquad (3.6a)$$

$$u_j^\pm = u_j - U_j^\pm . \qquad (3.6b)$$

These velocities are naturally related to the Favre and Reynolds averaged global material velocities U and V, respectively:

$$U = \frac{\overline{\varrho u}}{\overline{\varrho}} = \frac{\overline{\varrho c^+}}{\overline{\varrho}}\,\frac{\overline{\varrho c^+ u}}{\overline{c^+ \varrho}} + \frac{\overline{\varrho c^-}}{\overline{\varrho}}\,\frac{\overline{\varrho c^- u}}{\overline{c^- \varrho}} = C^+ U^+ + C^- U^- , \qquad (3.7a)$$

$$V = \overline{u} = \overline{c^+ u} + \overline{c^- u} = \alpha^+ U^+ + \alpha^- U^- + \overline{c^+ u^+} + \overline{c^- u^-} . \qquad (3.7b)$$

Thus the expression for the Reynolds averaged velocity contains turbulent mass flux terms for each fluid that are analogous to the single-fluid turbulent mass flux $\overline{u}$". These relations can be inverted to express $U^\pm$ as a function of U or V, and of the so-called drift or interpenetration velocity $\delta U = U^+ - U^-$:

$$U^\pm = \begin{cases} U \pm C^\mp \delta U , \\ V \pm \alpha^\mp \delta U - \overline{c^+ u^+} - \overline{c^- u^-} . \end{cases} \qquad (3.8)$$

The equations of mean motion are then rewritten using only $U^\pm$ to express the advection terms and introducing complementary corrective terms:

---

[2] It is also possible within this formalism to capture the contribution of an interfacial mixing zone at the *microscopic* scale by considering the second moments of the functions $c^\pm$; the "pure" portions of the fluids are given by the products with c^{+2} and c^{-2}, and the interface is given by $2c^+ c^-$, (we have of course $c^{+2} + 2c^+ c^- + c^{-2} = 1$).

[3] These equations take into account the deduced relation of mass conservation for any per mass quantity a:

$$\left(\tfrac{\partial}{\partial t}\left(c^\pm \varrho a\right) + \left(c^\pm \varrho a u_j\right)_{,j}\right) - c^\pm \left(\tfrac{\partial}{\partial t}\left(\varrho a\right) + \left(\varrho a u_j\right)_{,j}\right) = \left(\tfrac{\partial}{\partial t} c^\pm + \left(c^\pm\right)_{,j} u_j\right) \varrho a$$

$$= \left[\left(\tfrac{\partial}{\partial t}\left(c^\pm \varrho\right) + \left(c^\pm \varrho\right)_{,j} u_j\right) - c^\pm \left(\tfrac{\partial}{\partial t}\varrho + \left(\varrho\right)_{,j} u_j\right)\right] a$$

$$= \left[\left(\tfrac{\partial}{\partial t}\left(c^\pm \varrho\right) + \left(c^\pm \varrho u_j\right)_{,j}\right) - c^\pm \left(\tfrac{\partial}{\partial t}\varrho + \left(\varrho u_j\right)_{,j}\right)\right] a$$

$$= -\left(\phi_j^\pm\right)_{,j} a .$$

$$\frac{\partial}{\partial t}\left(\overline{c^\pm \varrho}\right) + \left(\overline{c^\pm \varrho \ U_j^\pm}\right)_{,j} = -\left(\overline{\phi_j^\pm}\right)_{,j}$$

$$\frac{\partial}{\partial t}\left(\overline{c^\pm \varrho U_i^\pm}\right) + \left(\overline{c^\pm \varrho U_i^\pm U_j^\pm}\right)_{,j} = -\left(\overline{c^\pm p}\right)_{,i} - \left(\overline{c^\pm \tau_{ij}}\right)_{,j} - \left(\overline{c^\pm \varrho u_i^\pm u_j^\pm}\right)_{,j} + \overline{c^\pm \varrho g_i}$$
$$+ \overline{(c^\pm)_{,i} p} + \overline{(c^\pm)_{,j} \tau_{ij}} - \overline{(\phi_j^\pm)_{,j} u_i}$$

$$\frac{\partial}{\partial t}\left(\overline{c^\pm \varrho f}\right) + \left(\overline{c^\pm \varrho f \ U_j^\pm}\right)_{,j} = -\left(\overline{c^\pm p u_j}\right)_{,j} - \left(\overline{c^\pm \tau_{ij} u_i}\right)_{,j} - \left(\overline{c^\pm \theta_j}\right)_{,j} - \left(\overline{c^\pm \varrho f u_j^\pm}\right)_{,j}$$
$$+ \overline{(c^\pm)_{,j} p u_j} + \overline{(c^\pm)_{,j} \tau_{ij} u_i} + \overline{(c^\pm)_{,j} \theta_j} - \overline{(\phi_j^\pm)_{,j} f}$$
$$+ \overline{c^\pm s} + \overline{c^\pm \varrho U_i^\pm g_i}$$

$$\frac{\partial}{\partial t}\left(\overline{c^\pm \varrho e}\right) + \left(\overline{c^\pm \varrho e \ U_j^\pm}\right)_{,j} = -\overline{c^\pm p(u_j)_{,j}} - \overline{c^\pm \tau_{ij}(u_i)_{,j}} - \left(\overline{c^\pm \theta_j}\right)_{,j} - \left(\overline{c^\pm \varrho e u_j^\pm}\right)_{,j}$$
$$+ \overline{(c^\pm)_{,j} \theta_j} - \overline{(\phi_j^\pm)_{,j} e} + \overline{c^\pm s}$$

$$\frac{\partial}{\partial t}\left(\overline{c^\pm \varrho \frac{u_i u_i}{2}}\right) + \left(\overline{c^\pm \varrho \frac{u_i u_i}{2} U_j^\pm}\right)_{,j} = -\overline{(c^\pm p)_{,i} u_i} - \overline{(c^\pm \tau_{ij})_{,j} u_i} - \left(\overline{c^\pm \varrho \frac{u_i u_i}{2} u_j^\pm}\right)_{,j}$$
$$+ \overline{(c^\pm)_{,j} p u_j} + \overline{(c^\pm)_{,j} \tau_{ij} u_i} - \overline{(\phi_j^\pm)_{,j} \frac{u_i u_i}{2}} + \overline{c^\pm \varrho U_i^\pm g_i}$$

$$(3.9)$$

where the mean per volume total energy of each fluid is expressed as:

$$\overline{c^{\pm}\varrho f} = \overline{c^{\pm}\varrho e} + \overline{c^{\pm}\varrho u^2}/2 = \overline{c^{\pm}\varrho e} + \overline{c^{\pm}\varrho}(U^{\pm})^2/2 + \overline{c^{\pm}\varrho(u^{\pm})^2}/2 \ . \tag{3.10}$$

Thus we obtain the Reynolds tensors for each fluid:

$$r_{ij}^{\pm} = \varrho u_i^{\pm} u_j^{\pm} \ , \tag{3.11}$$

whose evolution equation, which will not be discussed here, can be obtained as in the single-fluid case. We note that even after weighting with the volume fractions and after averaging, the Reynolds tensors for each fluid do not identify with the single-fluid Reynolds tensors: $\overline{c^{\pm}\varrho u_i'' u_j''} \neq \overline{c^{\pm}\varrho u_i^{\pm} u_j^{\pm}}$. The relationship between these quantities will be discussed in more detail in Sect. 7.3. The total kinetic and turbulent energies can be expressed as a function of their single-fluid analogs and of the drift velocity δU:

$$\overline{c^+\varrho}\,(U^+)^2/2 + \overline{c^-\varrho}\,(U^-)^2/2 = \overline{\varrho}\,U^2/2 + C^+ C^-\,\overline{\varrho}\,(\delta U)^2/2 \ , \tag{3.12a}$$

$$\overline{c^+\varrho\,(u^+)^2}/2 + \overline{c^-\varrho\,(u^-)^2}/2 = \overline{\varrho\,(u'')^2}/2 - C^+ C^-\,\overline{\varrho}\,(\delta U)^2/2 \ . \tag{3.12b}$$

Between the single-fluid and two-fluid approaches, we thus see the transfer of a portion of the turbulent energy in favor of kinetic energy. We will call this term the *directed per mass kinetic energy*, k_d, the remainder being called the *two-fluid turbulent kinetic energy*, k_b:

$$\left.\begin{array}{l} k_d = C^+ C^-\,(\delta U)^2/2 \ , \\ k_b = c^+ (u^+)^2/2 + c^- (u^-)^2/2 \end{array}\right\} \ \Rightarrow \ \tilde{k} = \tilde{k}_b + k_d \ . \tag{3.13}$$

As in (2.11), we balance the energies for each fluid:

Equation (3.14) — a single large multi-column balance equation. Reading left-to-right:

Left-hand side: Averaged values + Favre advections

$$\frac{\partial}{\partial t}\left(\ \overline{c^\pm \varrho}\ \right)+\left(\ \overline{c^\pm \varrho e}\ \ U_j^\pm\ \right)_{,j}$$

$$\frac{\partial}{\partial t}\left(\ \overline{c^\pm \varrho k^\pm}\ \right)+\left(\ \overline{c^\pm \varrho k^\pm}\ \ U_j^\pm\ \right)_{,j}$$

$$\frac{\partial}{\partial t}\left(\ \overline{c^\pm \varrho U_i^\pm U_i^\pm/2}\ \right)+\left(\ (\overline{c^\pm \varrho U_i^\pm U_i^\pm}/2)\,U_j^\pm\ \right)_{,j}$$

$$\frac{\partial}{\partial t}\left(\ \overline{c^\pm \varrho f}\ \right)+\left(\ \overline{c^\pm \varrho f}\ \ U_j^\pm\ \right)_{,j}$$

$=$

Mean motions for each fluid

	Sources	Sound and RT	Buoyancy	Anisobary	Dissipations
	$\overline{c^\pm s}$	$-\,\overline{\bar p\,c^\pm(u_i)}_{,i}$		$-\,\overline{c^\pm p'(\overline{u_i})}_{,i}$	$-\,\overline{c^\pm \tau_{ij}(U_i^\pm)}_{,j}\ -\ (\overline{c^\pm \theta_j})_{,j}$
		$-\,(\overline{\bar p\,c^\pm})_{,i}\,u_i^\pm$	$+\,\bar p\,\overline{(c^\pm)}_{,i}\,u_i^\pm$	$-\,(\overline{c^\pm p'})_{,i}\,u_i^\pm$	
	$\overline{c^\pm \varrho U_i^\pm g_i}$	$-\,(\overline{\bar p\,c^\pm})_{,i}\,U_i^\pm$	$+\,\bar p\,\overline{(c^\pm)}_{,i}\,U_i^\pm$	$-\,(\overline{c^\pm p'})_{,i}\,U_i^\pm$	$-\,(\overline{c^\pm \tau_{ij}})_{,j}\,U_i^\pm$
	$\overline{c^\pm s + c^\pm \varrho U_i^\pm g_i}$	$-\,(\overline{\bar p\,c^\pm}_{,i}\,u_i)$	$+\,\bar p\,\overline{(c^\pm)}_{,i}\,u_i$	$-\,(\overline{c^\pm p'\,u_i})_{,i}$	$-\,(\overline{c^\pm \tau_{ij}\,U_i})_{,j}\ -\ (\overline{c^\pm \theta_j})_{,j}$

$+$

Turbulent motions for each fluid

Noise	Pressure	Dissipation	Turbulent fluxes	Production
$-\,\overline{(c^\pm p')'(u_i)}_{,i}$		$-\,\overline{c^\pm \tau_{ij}(u_i^\pm)}_{,j}$	$-\,(\overline{c^\pm \varrho e\pm u_j^\pm})_{,j}$	
$+\,\overline{(c^\pm p')'(u_i)}_{,i}$	$-\,((\overline{c^\pm p'})'u_i^\pm)_{,i}$	$-\,(\overline{c^\pm \tau_{ij}})_{,j}\,u_i^\pm$	$-\,(\overline{c^\pm \varrho k\pm u_j^\pm})_{,j}$	$-\,\overline{c^\pm r_{ij}^\pm(U_i^\pm)}_{,j}$
	$-\,((\overline{c^\pm p')'u_i^\pm})_{,i}$	$-\,(\overline{c^\pm \tau_{ij}\,u_i})_{,j}$		$-\,(\overline{c^\pm r_{ij}^\pm})_{,j}\,U_i^\pm$
			$-\,(\overline{c^\pm \varrho f\pm u_j^\pm})_{,j}$	

$+$

Interfacial exchanges between fluids

Drag	Conduction	Interdiffusion
$+\,(\overline{c^\pm})_{,j}\,(p'\delta_{ij}+\tau_{ij})\,u_i^\pm$	$+\,(\overline{c^\pm})_{,j}\,\theta_j$	$-\,(\overline{\phi_j^\pm})_{,j}\,e$
$+\,(\overline{c^\pm})_{,j}\,(p'\delta_{ij}+\tau_{ij})\,U_i^\pm$		$-\,(\overline{\phi_j^\pm})_{,j}\,k^\pm$
	$-\,(\overline{\phi_j^\pm})_{,j}\,\theta_j$	$-\,(\overline{\phi_j^\pm})_{,j}\,U_i^\pm U_i^\pm/2\ -\ (\overline{\phi_j^\pm})_{,j}\,u_i^\pm U_i^\pm$
$+\,(\overline{c^\pm})_{,j}\,(p'\delta_{ij}+\tau_{ij})\,u_i$	$+\,(\overline{c^\pm})_{,j}\,\theta_j$	$-\,(\overline{\phi_j^\pm})_{,j}\,f$

$$(3.14)$$

Here, as for (2.11), the pressure was decomposed into mean and fluctuation, but in addition, the latter was itself separated into fluctuations *between the fluids* (so-called "anisobary" terms, or pressure difference terms at the interface, $\overline{c^{\pm}p'}$) and *within each fluid* (so-called "noise" terms $(c^{\pm}p')'$).

The interface terms represent energy exchanges *between fluids* and *between reservoirs* (internal, turbulent, or kinetic). We can separate the various contributions in a natural manner, because the terms $(c^{\pm})_{,j}$ and $(\phi_j^{\pm})_{,j}$, with opposite signs for the two fluids, are significant only in the vicinity of the *interface*, where the relevant mean velocities are $(U_i^{+}+U_i^{-})/2$ and $(U_i^{+}-U_i^{-})/2$. Noting the identity:

$$u_i^{\pm} = \frac{u_i^{+}+u_i^{-}}{2} \pm \frac{u_i^{+}-u_i^{-}}{2} = \frac{u_i^{+}+u_i^{-}}{2} \mp \frac{\delta U_i}{2} \, , \tag{3.15}$$

the drag terms in (3.14) are decomposed into exchanges between fluids and turbulent dissipations:

$$
\left[
\begin{array}{c}
\overline{\hspace{1cm}\text{Drag}\hspace{1cm}} \\
\text{Exchanges between fluids} \hspace{1cm} \text{Turbulent dissipations} \\[4pt]
\overline{(c^{\pm})_{,j}\,(p'\delta_{ij}+\tau_{ij})\,(u_i^{+}+u_i^{-})/2} \mp \overline{(c^{\pm})_{,j}\,(p'\delta_{ij}+\tau_{ij})}\,\delta U_i/2 \\
\overline{(c^{\pm})_{,j}\,(p'\delta_{ij}+\tau_{ij})}\,(U_i^{+}+U_i^{-})/2 \pm \overline{(c^{\pm})_{,j}\,(p'\delta_{ij}+\tau_{ij})}\,\delta U_i/2 \\
\overline{(c^{\pm})_{,j}\,(p'\delta_{ij}+\tau_{ij})\,u_i}
\end{array}
\right] . \tag{3.16}
$$

We see that the momentum exchange between fluids, which is related to the drag terms in $\overline{(c^{\pm})_{,j}\,(p'\delta_{ij}+\tau_{ij})}$ is responsible not only for transfers of mean kinetic energies and turbulent energies between fluids, but also for a conversion of mean kinetic energies into turbulent energies. This latter represents the dissipation mechanism of directed kinetic energy into turbulent energy.

3.2 An Elementary Two-Fluid Closure: AWE's Model

In contrast to the single-fluid approach, where most two-equation turbulence models are practically equivalent to the k–ε model, the two-fluid approach has led to a wide variety of models, even for elementary situations [83–86]. The model to be discussed here is that proposed by D.L. Youngs of the AWE for the study of instabilities in laser targets [33]. The details of the approximations that lead to the closures have not been completely disclosed, so we shall present what seems to be the probable course. The notation was adapted to be consistent with the rest of the present study.

1. The fluids are assumed to be immiscible (and thus without interdiffusion) and not heat-conducting:

$$c^{\pm} \overset{m}{=} 0 \text{ or } 1 \, , \quad \text{with} \quad \phi \overset{m}{=} \theta \overset{m}{=} 0 \, . \tag{3.17}$$

2. The acoustic turbulent energies are negligible *except in the presence of shocks* (low turbulent Mach number):

$$\overline{(c^{\pm}p')'(u_i)_{,i}} \stackrel{\mathrm{m}}{=} 0 . \tag{3.18}$$

3. The mean tensors of viscous stresses are negligible, except in the presence of shocks (high turbulent Reynolds number):

$$\overline{c^{\pm}\tau_{ij}} \stackrel{\mathrm{m}}{=} 0 . \tag{3.19}$$

4. To permit the expression of various correlations with the density, particularly for the variables of state of the fluids, the density fluctuations are negligible *in each fluid*:

$$c^{\pm}(\varrho - \varrho^{\pm}) \stackrel{\mathrm{m}}{=} 0 , \tag{3.20}$$

except when calculating the Reynolds averaged velocities (see next item).

5. The Reynolds averaged velocities (global and for each fluid) are obtained by first-gradient closure of the turbulent mass fluxes for each fluid (with no adiabatic corrections, see E):

$$V_i^{\pm} = \overline{c^{\pm}u_i}/\overline{c^{\pm}} = U_i^{\pm} + \overline{c^{\pm}u_i^{\pm}}/\overline{c^{\pm}} \stackrel{\mathrm{m}}{=} U_i^{\pm} + D_t \frac{(\varrho^{\pm})_{,i}}{\varrho^{\pm}} , \tag{3.21}$$

where D_t is the turbulent diffusion coefficient defined below in (3.33c). This approach is analogous to that used in the single-fluid case for the "Rayleigh–Taylor" production term in (2.17) and has some of the disadvantages discussed in Sect. 7.2. We deduce the expression for the overall Reynolds averaged velocity:

$$V_i = \alpha^+ V_i^+ + \alpha^- V_i^-$$
$$\stackrel{\mathrm{m}}{=} \alpha^+ U_i^+ + \alpha^- U_i^- + D_t \left(\alpha^+ \frac{(\varrho^+)_{,i}}{\varrho^+} + \alpha^- \frac{(\varrho^-)_{,i}}{\varrho^-} \right) , \tag{3.22}$$

which thus contains a complementary diffusive term compared with the definition used in most two-fluid models [83–86].

6. The pressure difference between the fluids, although small, is not disregarded in the internal energy equations (two-pressure model). Thus for the internal energies in (3.14) we can write:

$$\overline{\bar{p}c^{\pm}(u_i)_{,i}} + \overline{c^{\pm}p'(\overline{u_i})_{,i}} \stackrel{\mathrm{m}}{=} \beta^{\pm} P^{\pm}(V_i)_{,i} , \tag{3.23}$$

where:

$$\beta^{\pm} = \frac{\alpha^{\pm}/(\varrho^{\pm}c_s^{2\pm})}{\alpha^+/(\varrho^+c_s^{2+}) + \alpha^-/(\varrho^-c_s^{2-})} . \tag{3.24}$$

The $\beta^{\pm}$ are the distribution coefficients of the pressure work given as functions of the relative compressibilities of the fluids.

7. The pressure difference between the fluids is disregarded in the momentum equations (3.9):

$$\overline{c^\pm(p)_{,i}} \overset{\text{m}}{=} \alpha^\pm P_{,i} \, , \tag{3.25}$$

with:

$$P = \alpha^+ P^+ + \alpha^- P^- \, , \tag{3.26}$$

from energy conservation.

8. The pressure difference between the fluids is determined by a supplementary empirical equation for the relaxation of volume fractions:

$$\tfrac{\partial}{\partial t}\alpha^\pm + (\alpha^\pm)_{,j} V_j \overset{\text{m}}{=} - \left(\alpha^\pm(V_j^\pm - V_j)\right)_{,j} \, . \tag{3.27}$$

The addition of an equation in $\alpha^\pm$ is a common approach in two-fluid models [83–86], but the relaxation term is unusual. The underlying dissipative mechanism appears to be turbulent diffusion, which is included in the definition of the $V^\pm$.

All these hypotheses are classical for two-fluid flows and are generally fairly well verified. To these are added the following turbulent closures, which are specific to AWE's two-fluid model and thus are less well established:

9. It is assumed that the Atwood number is small and that the integral length scale of turbulence is larger than the characteristic sizes of the mixture structures. Then an effective mean fluid for the turbulence can assumed, thus reducing the two-turbulence statistical equations to a one-turbulence model:

$$\overline{c^\pm \varrho u_i^\pm u_j^\pm} \overset{\text{m}}{=} \overline{c^\pm \varrho} \; \overline{u_i^\pm u_j^\pm} = C^\pm R_{ij} \, , \tag{3.28}$$

(recalling that $C^\pm$ are the mean mass fractions).

10. In this one-turbulence approximation, the evolution equation for the two-fluid turbulent kinetic energy, $\tilde{k}_b$ defined in (3.13) is assumed identical to the equation of $\tilde{k}$ in a single-fluid model. The evolution of $\tilde{k}_b$ is then described by the modeled equation (2.20), where only the production terms are modified: this equation is different from that obtained by summing the equations in $\overline{c^\pm \varrho k^\pm}$ in (3.14) (in the latter case we find complementary terms to the turbulent fluxes, see Sect. 7.3). By simple conservation of energy, the total production is written as:

$$- R_{ij} \left(C^+(U_i^+)_{,j} + C^-(U_i^-)_{,j} \right) + D_i(U_i^+ - U_i^-)$$
$$= - R_{ij}(U_i)_{,j} + (D_i \pm (C^\pm)_{,j} R_{ij}) \, \delta U_i \, , \tag{3.29}$$

which includes the work of the drag force, $D_i = \mp \overline{(c^\pm)_{,j} \, (p'\delta_{ij} + \tau_{ij})}$, in place of the "Rayleigh–Taylor" production term in (2.20).

11. The algebraic closure of the Reynolds tensor is identical to (2.14) for the k-ε model, with the same definition of the turbulent viscosity ν_t.

12. The drag is closed according to an empirical expression valid in the Newtonian regime (strong Reynolds and small Mach numbers) with a correction for the added mass (with a coefficient of $1/2$, adjusted for spheres) [83–88, §I.3.3]:

$$D_i \pm (C^{\pm})_{,j} R_{ij} = D_i^*$$

$$\stackrel{\mathrm{m}}{=} C_d \, \overline{\varrho} \, \frac{\alpha^+ \alpha^-}{\lambda_d} \, \| \, \delta \boldsymbol{U} - \boldsymbol{W} \, \| \, (\delta U_i - W_i)$$

$$+ \frac{1}{2} \overline{\varrho} \, \alpha^+ \alpha^- \left(\tfrac{\partial}{\partial t}(U_i^+ - U_i^-) + U_j^+ (U_i^+)_{,j} - U_j^- (U_i^-)_{,j} \right) , \quad (3.30)$$

where λ_d is a dimension of the fluid structures that is characteristic of the drag, C_d is a drag coefficient, and $\boldsymbol{W}$ is the velocity of turbulent dispersion:

$$W_i = -D_t \left(\frac{(\alpha^+ \varrho^+)_{,i}}{\alpha^+ \varrho^+} - \frac{(\alpha^- \varrho^-)_{,i}}{\alpha^- \varrho^-} \right) , \quad (3.31)$$

according to a classical formulation for dispersed flows [89, 90]. As for the turbulent fluxes, $\boldsymbol{W}$ can be subjected to a limitation to guarantee realizability (see item 3 in Sect. 2.2). The important feature of this formulation is the presence of the corrective term $\mp (C^{\pm})_{,j} R_{ij}$, which is justified [33] by the need to obtain $\delta U = W$ under the sole influence of turbulent diffusion and in the absence of pressure gradients and inertial effects. This correction is to be compared with the work of drag as given in (3.29).

13. An evolution equation for λ_d is constructed in an entirely heuristic manner [29, 30]:

$$\tfrac{\partial}{\partial t} \lambda_d + \left(\alpha^- U_i^+ + \alpha^+ U_i^- + \alpha^+ D_t \frac{(\varrho^+)_{,i}}{\varrho^+} + \alpha^- D_t \frac{(\varrho^-)_{,i}}{\varrho^-} \right) (\lambda_d)_{,i}$$

$$= [D_t (\lambda_d)_{,i}]_{,i} + [\, (1 + C_\lambda)(V_i)_{,i}/3$$

$$- C_\lambda \, n_i n_j ((V_i)_{,j} + (V_j)_{,i})/2 \,] \lambda_d + \sqrt{\frac{2 \overline{\varrho}}{\varrho^+ + \varrho^-}} \, n_i \delta U_i , \quad (3.32)$$

where n_i is the unit vector directed along $\mp (\alpha^{\pm})_{,i}$ and C_λ is a phenomenological constant. We note that:

- the advection velocity of λ_d is a mean velocity weighted by the inverse volume fractions (λ_d follows more the least abundant fluid) with the addition of a correction for turbulent dispersion;
- the homogeneous next-to-last term describes the production by the deformation of structures induced by the mean motion; and
- the last production term, which is heuristic and inhomogeneous, is proportional to the interpenetration velocity along the mixing gradient, and was chosen to provide a length λ_d that is approximately uniform in a mixing layer.

14. The turbulent coefficients of viscosity, diffusion, and dissipation are expressed in terms of the length λ_d as:

$$\nu_t = C_\mu \sqrt{\tilde{k}_b}\, \lambda_i \,, \tag{3.33a}$$

$$D_t = 2\nu_t \,, \tag{3.33b}$$

$$\tilde{\varepsilon} = \tilde{k}_b^{\,3/2}/\lambda_i \,, \tag{3.33c}$$

$$\lambda_i = (C_i/C_\mu)\,\lambda_d \,. \tag{3.33d}$$

With the notations of the k–ε model in Sect. 2.2, this is equivalent to taking the Schmidt–Prandtl numbers, σ, all equal to .5.

15. As in the k–ε model, the physical and numerical treatment of shocks [91] consists of adding to the pressure a single artificial viscosity stress in the equations of momentum and internal energy. The artificial viscosity is of the Wilkins type calculated from the *volume averaged* velocity gradient $V_{i,j}$ so as to remain compatible with the work of the pressure forces in the internal energy equations. In rare cases with high Atwood numbers, instabilities can occur that require the use of different artificial viscosities for the two fluids (see Sect. 8.3).

16. The recommended numerical constants are chosen to reconstruct the self-similar Kelvin–Helmholtz and Rayleigh–Taylor mixing instabilities in the limit of zero Atwood number:[4]

Table 3.1. Recommended constants for AWE's two-fluid model [33]

C_μ	C_d	C_i	C_λ
.09	20	.105	1

To summarize, the equations of AWE's two-fluid model are [33]:

$$
\begin{cases}
\frac{\partial}{\partial t}(\alpha^\pm) + (\alpha^\pm \quad V_j^\pm)_{,j} \overset{m}{=} -\left(\alpha^\pm(V_j^\pm - V_j)\right)_{,j}, \\[4pt]
\frac{\partial}{\partial t}(\alpha^\pm \varrho^\pm) + (\alpha^\pm \varrho^\pm \, U_j^\pm)_{,j} = 0, \\[4pt]
\frac{\partial}{\partial t}(\alpha^\pm \varrho^\pm U_i^\pm) + (\alpha^\pm \varrho^\pm U_i^\pm U_j^\pm)_{,j} = -C^\pm(R_{ij})_{,j} - \alpha^\pm(P)_{,i} \quad \mp D_i^* \quad + \alpha^\pm \varrho^\pm g_i, \\[4pt]
\frac{\partial}{\partial t}(\alpha^\pm \varrho^\pm E^\pm) + (\alpha^\pm \varrho^\pm E^\pm U_j^\pm)_{,j} \overset{m}{=} -(\Phi_j^{e\pm})_{,j} \quad -\beta^\pm P^\pm(V_i)_{,i} + \alpha^\pm \varrho^\pm \tilde{\varepsilon} + \overline{c^\pm s}, \\[4pt]
\frac{\partial}{\partial t}(\overline{\varrho}\, \tilde{k}_b) + (\overline{\varrho}\, \tilde{k}_b \quad U_j)_{,j} \overset{m}{=} -(\Phi_j^k)_{,j} \quad + \pi_{KH} + \pi_D \quad - \overline{\varrho}\tilde{\varepsilon}, \\[4pt]
\frac{\partial}{\partial t}(\lambda_d) + (\lambda_d)_{,j} \quad V_j^\lambda \overset{m}{=} -(\Phi_j^\lambda)_{,j} \quad + \pi_{KH}^\lambda + \pi_D^\lambda,
\end{cases} \tag{3.34}
$$

where the turbulent fluxes are closed by:

---

[4] D.L. Youngs [33] uses the notations K and L for our k_b and λ_d, and the coefficients c_1, c_2, and c_3 for C_d, C_i and C_λ, respectively.

$$\Phi_j^a \overset{m}{=} -\overline{\varrho}\, D_t\, (\tilde{a})_{,j}\ , \tag{3.35a}$$

$$\Phi_j^{e\pm} \overset{m}{=} -\alpha^{\pm}\overline{\varrho}\, D_t\, (E^{\pm})_{,j}\ , \tag{3.35b}$$

$$R_{ij} \overset{m}{=} \overline{\varrho}\,\frac{2\widetilde{k}_b}{3}\,\delta_{ij} - \overline{\varrho}\,\nu_t\left[\,(U_i)_{,j}+(U_j)_{,i}\,\right]\ , \tag{3.35c}$$

and the V, $V^{\pm}$, $\beta^{\pm}$, D^*, $\pi_{KH}+\pi_D$, $\pi_{KH}^{\lambda}+\pi_D^{\lambda}$ and V^{λ} are closed as in (3.22), (3.21), (3.24), (3.30), (3.29), (3.32), and (3.32), respectively.

Summary of Part I

This first part has provided the reference framework of the two basic models found in the literature, DAM's k–ε and AWE's two-fluid, which embody the two main approaches used for gravitationnally driven turbulent mixing flows.

To facilitate the comparison between the two approaches carried out over this work, common notations have been used as much as possible and a common averaging procedure has been applied (ensemble averaging). At this early stage, some fundamental differences appear between single- and two-fluid equations which have been highlighted.

Most noticeable is the turbulent energy which involves a *directed energy* contribution which is separated by the two-fluid approach, as given in (3.13). This energy component, and the energy transfer paths involving it, will play a crucial role in the next chapters.

In both modeling routes, two production processes of turbulent kinetic energy (and directed energy) are present. The first is the usual work of Reynolds stresses through shear of the mean flow. The second takes noticeably different forms in the two models: enthalpic or "Rayleigh–Taylor" production for the single fluid model (2.17), and drag work for the two-fluid model (3.16).

One complementary equation is added in both models, respectively on ε or λ_d, with specific production terms. Although these quantities are very different in meaning and form, they will be related to the same physical observations. They thus play somewhat similar roles in closing the models.

Part II

"0D" Reduction of Experiments and Models
in Mixing Instabilities

4

Mixing Instabilities in Developed Regime: Phenomenology of Energy Balances

4.1 Review of Different Instability Types and Regimes

The three major types of "ideal" instabilities studied here are [92]:

- Kelvin–Helmholtz (KH): two fluids, possibly identical, separated by an initially plane surface, are each driven by different uniform velocities tangent to the plane;
- Rayleigh–Taylor (RT) [93–97]: two fluids at rest, of different densities, separated by an initially plane surface, are subjected to a gravity field perpendicular to the plane and oriented from heavy to light;
- Richtmyer–Meshkov (RM) [97]: two compressible fluids, of different impedances, separated by an initially plane surface, are set in motion by the passage of a shock propagating along the normal to the plane.

These three cases of flows (1D) are unstable with regard to small perturbations (2D or 3D) of the initial conditions: interface distortions, nonuniform velocity fields, etc.

Detailed analyses of these instabilities will not be undertaken here, but can be found in many publications [92–97]. We will recall only that four successive evolution regimes have been identified for these three types of instabilities:

1. linear: the perturbations are of small amplitude, and their evolution equations are linear; decomposition into eigenmodes (in general, as the result of Fourier analysis) leads to an analytic solution where the unstable modes grow exponentially (linearly for RM).
2. weakly nonlinear: the perturbations are of sufficient amplitude so that the evolution equations have nonlinear terms that are non-negligible, but small enough to be handled using perturbative methods; the evolution of eigenmodes is not strictly exponential, and mode and frequency-mixing effects appear.

A. Llor: *Statistical Hydrodynamic Models for Developed Mixing Instability Flows*, Lect. Notes Phys. **681**, 33–45 (2005)
www.springerlink.com

3. nonlinear or transient: the perturbations are of large amplitude and the nonlinear terms have become significant in the equations; evolution, which is more of an algebraic type, can no longer be described by superimposing eigenmodes, but certain characteristics of the initial perturbations are preserved, in particular the wavelength in the single-mode case.

4. developed turbulent: the perturbations have become large and chaotic, because the nonlinear terms dominate the evolution equations; there is now a certain *"loss of memory"* of the initial conditions, and the description of evolution must rely on turbulence modeling: statistical, spectral, large-scale simulation (LES), or direct simulation (DNS).

The discussion to follow will be concerned only with *statistical models* describing the *final regime*. In many applications, this is the dominant regime in terms of duration and effects.

In the case of incompressible homogeneous fluids with a constant driving term (shear in KH, gravity in RT), the flow in developed turbulent regime is self-similar.[1] Thus the "loss of memory" associated with this regime erases all the characteristic quantities of the flow, leaving as the only descriptive parameters time and the energy source. These two values are sufficient to give the general expressions for the mixing zone characteristics: the thickness $L(t)$ behaves as t for KH, t^2 for RT, and t^{n_0} where $n_0 \approx 1/3$ for RM (as shown in Table 4.1, p. 38). Physical and numerical experiments confirm these behaviors and allow an estimation of the universal coefficients associated with these laws.

4.2 "0D" Energy Balance in Self-Similar, Incompressible, Developed Regime at Zero Atwood Number

The growth of the mixing zone, as given by the law $L(t)$, globally describes the relative displacement of the fluids, and it is thus associated with two characteristic energies of the mixing zone: the *time integrated input* energy (from a kinetic or potential source for KH and RT instabilities, respectively) and the *directed* kinetic energy due to the mean velocity acquired by the fluids. Therefore, the knowledge of $L(t)$ provides two of the basic terms in the energy balance of a turbulent mixing layer. Borrowing from earlier methods [59] that were revived in part by Mikaelian for mixing layers [98,99] and have been used implicitly in several studies at the DAM [26,100–103] and elsewhere [104–109], a "0D" analysis for the energy balances can be obtained by further introducing

---

[1] We recall that a Richtmyer–Meshkov instability, although *initiated by a passing shock*, actually involves an *incompressible description of its evolution* (in the absence of any effect that could cause a variation in the densities of the fluids after the shock passage, as for example with certain combinations of impedances which result in an expansion wave).

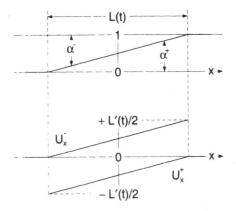

Fig. 4.1. Approximate profiles of volume fractions and velocities in a mixing zone, used for estimating various quantities of "0D" models

some simple, but realistic, approximations for the profiles of the main quantities in the mixing zone. The other more refined characteristics of turbulent mixing layers (detailed profiles, anisotropies, etc.) can be equally important in certain situations, but as we shall see in Chap. 7, they are marginal in many DAM applications.

As shown in Fig. 4.1, we shall assume here that the profiles of the volume fractions $\alpha^{\pm}$ of the two mixing fluids, indexed by $+$ and $-$, are linear:

$$\alpha^{\pm}(x) = \frac{1}{2} \pm \frac{x}{L} \ . \tag{4.1}$$

The mixing zone is therefore symmetric, which is rigorous only in the limit of fluids with identical properties. The analysis presented here will thus be valid only in the limit of zero Atwood numbers. Experimentally, the hypothesis of a linear profile has been fairly well verified, except at the edges of the mixing zones, where there is some rounding [31, 32, 110–124]. This could introduce some bias into the relation between L and the mean slope of α, but the thickness is usually obtained by a second moment method that corrects this effect:[2]

$$L = 6 \int_{-\infty}^{+\infty} \alpha^{+}(x)\,\alpha^{-}(x)\,\mathrm{d}x \ . \tag{4.2}$$

Returning to the two-fluid statistical definitions given in Chap. 3, we shall use the fluid velocities $\boldsymbol{U}^{\pm}$ to estimate the mean kinetic energies. With the assumptions of uniform densities, and of incompressible and immiscible fluids, the Reynolds and Favre averaged velocities, $\boldsymbol{V}$ and $\boldsymbol{U}$, and the turbulent

---

[2] Ⓤ Many authors [33, 120, 121] replace the coefficient 6 by 6.6 in this formula in order to take into account the slight smearing of the TMZ edges observed in experiments. This correction can be safely neglected considering the approximations used in the present study.

mass fluxes for each fluid $\overline{c^\pm u^\pm}$, vanish in the limit of zero Atwood number. According to (3.8), the velocity profiles of the fluids are written as:

$$U_x^\pm(x) = \pm \alpha^\mp(x)\delta U_x(x) \ . \tag{4.3}$$

Inserting these expressions and the linear profiles of the $\alpha^\pm$ into the mass conservation equations of the two fluids (see (3.9) in Sect. 3.1), we find by integrating in x that δU_x is *uniform* over the mixing zone. In addition, the evolution of the mixing zone edges is given by:

$$\frac{\mathrm{d}L}{\mathrm{d}t} = L' = U_x^-(+L/2) - U_x^+(-L/2) \ , \tag{4.4}$$

from which we finally obtain:

$$U_x^\pm(x) = -\left(\frac{x}{L} \mp \frac{1}{2}\right)\delta U_x \quad \text{and} \quad L' = -2\,\delta U_x \ , \tag{4.5}$$

($\delta U_x < 0$ with the axis choice adopted and shown in Fig. 4.1). The growth rate of the mixing zone thus makes it possible to express the two components, *directed* and *mean*, of the total per mass kinetic energy given in (3.13). The mean per mass values for the mixing zones, K_D and K_M are then given by:[3]

$$K_D = \frac{\int_{-L/2}^{+L/2} \overline{\varrho} k_d \, \mathrm{d}x}{\int_{-L/2}^{+L/2} \overline{\varrho} \, \mathrm{d}x} \approx \frac{1}{L}\int_{-L/2}^{+L/2} C^+ C^- (\delta U_x)^2 \! /_2 \, \mathrm{d}x$$

$$\approx \frac{1}{12}\,(\delta U_x)^2 = \frac{1}{48}\,(L')^2 \ , \tag{4.6a}$$

$$K_M = \frac{\int_{-L/2}^{+L/2} \overline{\varrho} (U_x)^2 \! /_2 \, \mathrm{d}x}{\int_{-L/2}^{+L/2} \overline{\varrho} \, \mathrm{d}x} \approx -\frac{1}{L}\int_{-L/2}^{+L/2} (\alpha^+ - C^+)(\alpha^- - C^-)(\delta U_x)^2 \! /_2 \, \mathrm{d}x$$

$$\approx \frac{1}{15}\,\mathcal{A}^2(\delta U_x)^2 = \frac{4\,\mathcal{A}^2}{5}\,K_D. \tag{4.6b}$$

These expressions are valid regardless of the type of self-similar instability under consideration or the nature of the fluid mixture (heterogeneous or intimate), since the value of δU_x depends only on the evolution law of $L(t)$. They show that the mean kinetic energy, K_M, represents only an asymptotically

---

[3] The approximations in these expressions of K_D and K_M are valid in the limit $\mathcal{A} \to 0$, where $\overline{\varrho}$ can be considered uniform. We also note that:

$$\alpha^\pm - C^\pm = \alpha^\pm\left(1 - \frac{\varrho^\pm}{\overline{\varrho}}\right) = \mp 2\alpha^+ \alpha^- \frac{\mathcal{A}}{1 + \mathcal{A}(\alpha^+ - \alpha^-)}$$

small portion of the directed kinetic energy K_D as $\mathcal{A} \to 0$. Thus *we will disregard K_M for the rest of this study.* The values of K_D in the three cases of KH, RT, and RM are given in Table 4.1.

Now we examine the time integrated input energy, K_I, which is the other energy balance term imposed by $L(t)$. In the RT case, given the profile of $\alpha^\pm$, we can calculate the variation of the mean per mass gravitational energy of the system, representing the balance of the input energy and energy dissipated to turbulent or internal energies:

$$K_I^{\boxed{RT}} = \frac{-2\,\Gamma}{(\varrho^+ + \varrho^-)\,L} \left(\int_{-L/2}^{+L/2} (\alpha^+(x)\varrho^+ + \alpha^-(x)\varrho^-)\,x\,\mathrm{d}x \right.$$
$$\left. - \int_{-L/2}^{0} \varrho^-\,x\,\mathrm{d}x - \int_{0}^{+L/2} \varrho^+\,x\,\mathrm{d}x \right) , \quad (4.7)$$

where, with the axis choice shown in Fig. 4.1, the acceleration field is written as $g_x = -\Gamma$, with $\Gamma > 0$. In the limit $\mathcal{A} \to 0$, where the mean density in the mixing zone is fairly constant and is given by $(\varrho^+ + \varrho^-)/2$, the result of the integration is given in Table 4.1. In the KH case, *an additional assumption on the transverse velocity profile in the mixing zone* is required to determine the input energy balance. Assuming a linear profile over the *same* thickness L, defined by the profile of $U^\pm(x)$, we obtain:

$$K_I^{\boxed{KH}} = \frac{-(\Delta U_y/2)^2}{(\varrho^+ + \varrho^-)\,L} \left(\int_{-L/2}^{+L/2} (\alpha^+(x)\varrho^+ + \alpha^-(x)\varrho^-) \left(\frac{2x}{L}\right)^2 \mathrm{d}x \right.$$
$$\left. - \int_{-L/2}^{0} \varrho^-\,\mathrm{d}x - \int_{0}^{+L/2} \varrho^+\,\mathrm{d}x \right) , \quad (4.8)$$

where ΔU_y is the difference of the fluid velocities at infinity. The result is shown in Table 4.1. In the RM case, the total input energy during the passage of the initial shock is a constant of the motion, since there is no production mechanism during self-similar growth: thus K_I decreases as $1/L$.

4.3 "0D" Turbulent Energy and Structure in Self-Similar, Incompressible, Developed Regime at Zero Atwood Number

According to Sect. 4.2 above, the directed kinetic energy is always a small fraction of the input energy, K_I: most of K_I is converted into turbulent kinetic energy, K, which in turn is dissipated. A characterization of the energy of turbulent mixing zones must therefore include data on K obtained experimentally or numerically. For purposes of comparison with the one- and two-fluid

Table 4.1. Basic geometric and (per mass) energy characteristics of self-similar turbulent mixing zones generated by Kelvin–Helmholtz, Rayleigh–Taylor, and Richtmyer–Meshkov instabilities (averaged estimates using the "0D" approach at $\mathcal{A} \to 0$ for linear density and velocity profiles as in Fig. 4.1). The von Kármán number characterizes the relative size of the turbulent structures and is related to the energy dissipation. ① A recent comparative study of experimental and numerical results from various groups [123] points towards a smaller value of the mean of the observed distribution of $\mathcal{Y}_0$ values. This correction can be safely neglected considering the approximations used in the present study

Type of Instability	KH	RT	RM
Observed "universal" constant (experiment or DNS, at $\mathcal{A} \to 0$)	$\mathcal{X}_0 \approx .1$	$\mathcal{Y}_0 \approx .12$	$n_0 \approx .3$
L: (observed) thickness of mixing zone	$\mathcal{X}_0 \times \Delta U_y t$	$\mathcal{Y}_0 \times A\Gamma t^2$	$L_0 \left(\dfrac{t}{t_0} \right)^{n_0}$
K_I: total input energy in mixing zone	$\dfrac{1}{12} \times (\Delta U_y)^2$	$\dfrac{\mathcal{Y}_0}{12} \times (A\Gamma t)^2$	$K_{I0} \left(\dfrac{t}{t_0} \right)^{-n_0}$
K_D: directed kinetic energy	$\dfrac{\mathcal{X}_0^2}{48} \times (\Delta U_y)^2$	$\dfrac{\mathcal{Y}_0^2}{12} \times (A\Gamma t)^2$	$\dfrac{n_0^2}{48} \times \left(\dfrac{L_0}{t_0} \right)^2 \left(\dfrac{t}{t_0} \right)^{-2(1-n_0)}$
K_D/K_I	$\dfrac{\mathcal{X}_0^2}{4}$	$\mathcal{Y}_0$	$\dfrac{n_0^2}{48} \times \dfrac{L_0^2}{K_{I0} t_0^2} \left(\dfrac{t}{t_0} \right)^{-(2-3n_0)}$

Table 4.1. continued

Type of Instability	KH	RT	RM
K: single-fluid turbulent energy	$\approx \left\{ \begin{array}{l} .29 \times K_I \\ 116 \times K_D \end{array} \right.$	$\approx \left\{ \begin{array}{l} .48 \times K_I \\ 4 \times K_D \end{array} \right.$	$\approx 58 \times K_D$
K_B: two-fluid turbulent energy		$\approx \left\{ \begin{array}{l} .36 \times K_I \\ 3 \times K_D \end{array} \right.$	
κ: single-fluid von Kármán number	$\approx .63$	$\approx .13$	$\approx .3$
κ_b: two-fluid von Kármán number		$\approx .09$	
References experimental: numerical (3D):	[110–117] [125–127]	[31,32,118–123] [33,128–130] [122,123]	[120,121,124] [131,132] [33]

models, we included in Table 4.1 both K and $K_B = K - K_D$, the mean *two-fluid* turbulent energy defined in (3.13), which are given by their ratios to K_I and K_D. We note that in the KH case, these quantities are well known from experiments [110–117], while we were able to find only one publication [33] on a numerical simulation for the RM case[4] (the RT case being an intermediate situation). Thus these quantities are not known with the precision that might be desired, although it is sufficient for the present study.

The energy balance is now fully defined, the remainder $K_I - K$ being dissipated as heat. However, it is useful to include three quantities that are characteristic of this dissipation (and thus redundant but important for modeling): the mean dissipation $E = \langle \varepsilon \rangle$, obtained from the relation:

$$L E = \frac{\mathrm{d}}{\mathrm{d}t}(L K_I - L K) , \qquad (4.9)$$

the integral turbulence length scale, Λ_i:[5]

$$\Lambda_i = \frac{K^{3/2}}{E} , \qquad (4.10)$$

and the "von Kármán number," κ:

$$\boxed{\kappa = \frac{\Lambda_i}{L} = \frac{K^{3/2}}{E L} .} \qquad (4.11)$$

The dissipation, E, is a quantity defined in an absolute manner by the balance relation (4.9), and does not as such bring any new information. By contrast, the definition of the integral length scale, which estimates the characteristic size of the largest turbulent eddies, makes the implicit hypothesis of spectral quasi-equilibrium of the turbulent energy. Experiments [122] (and to lesser extent numerical simulations) indicate that this condition is safely fulfilled for

---

[4] Shock tube measurements of velocity fluctuations at the DAM in Vaujours are currently being reinterpreted by J.–F. Haas, G. Rodriguez, F. Poggi, and X. Rogue. This work has confirmed the order of magnitude of the von Kármán number obtained by direct simulation for the RM case.

[5] ⓘ Various authors in the turbulence community consider $k^{3/2}/\varepsilon$ to be only proportional the integral length scale in general. The factors, close to 1, that they introduce in their definitions often depend on constants of specific models and may vary according to flow situations. However, many standard textbooks on turbulence do define $k^{3/2}/\varepsilon$ as the integral length scale under the more or less explicit assumptions of spectral quasi-equilibrium (experimentally observed in RT, see footnote 6, p. 41) and universality of the Kolmogorov constant (also experimentally verified in many different flows). This ensures that *very different flow situations can be analyzed with this common yardstick*. The present study critically hinges on this property when comparing KH, RT and RM instabilities in order to reveal the distinctive features of RT flows. See also footnote 8, p. 42.

the self-similar KH, RT, and RM flows of the present study.[6] The dimensionless geometric parameter κ thus indicates how many of the largest turbulent eddies are contained within the mixing zone width. A similarly defined ratio κ appears as the von Kármán *constant* in the analysis of turbulent boundary layers [59, §5.2], [134, §17.1.4], which greatly resemble the Kelvin–Helmholtz instability. Thus, by extension, we have adopted here the name von Kármán *number* for κ.

The von Kármán number κ represents, in a derived form, the dissipation in the mixing layer, but its specific importance stems from being identified as a *Knudsen number* that characterizes the turbulent transport, as has been previously noted [59, §2.3]. Indeed the turbulent diffusion is associated with a mean free path of the order of Λ_i, and the value of κ thus represents the first criterion for the validity of the turbulent flux closures in the models. For example, the coefficients of the k–ε model were determined so as to capture various experimental parameters, including the von Kármán constant [62,69–76].[7]

For the three self-similar instabilities being studied, equations (4.9) through (4.11) allow the von Kármán number to be expressed as a function of the characteristic growth coefficients and the ratios K/K_I or K/K_D:

$$
\kappa = \begin{cases}
\dfrac{(K/K_I)^{3/2}}{1-(K/K_I)} \times \dfrac{\sqrt{K_I/L}}{\dfrac{\mathrm{d}\,LK_I}{LK_I\,\mathrm{d}\,t}} = \dfrac{(K/K_I)^{3/2}}{1-(K/K_I)} \times \begin{cases} \dfrac{1}{2\sqrt{3}\mathcal{X}_0} & \text{KH}, \\[2ex] \dfrac{1}{8\sqrt{3}\mathcal{Y}_0} & \text{RT}, \end{cases} \\[6ex]
\sqrt{\dfrac{K}{K_D}} \times \dfrac{\sqrt{K_D/L}}{-\dfrac{\mathrm{d}\,LK_D}{LK_D\,\mathrm{d}\,t}} \;=\; \sqrt{\dfrac{K}{K_D}} \times \dfrac{n_0}{4\sqrt{3}\,(2-3n_0)} & \text{RM}.
\end{cases}
$$

$$(4.12)$$

The numerical values of these expressions are given in Table 4.1, p. 38. We note here that a precise measurement of the ratio K/K_I is important in the KH and RT cases, because the von Kármán number diverges as $K \to K_I$.

---

[6] Ⓤ Despite the accumulation of unambiguous evidence, even recent and of high quality [133], the existence of spectral quasi-equilibrium in the RT flow is still questioned sometimes: the basic argument is that the t^2 TMZ growth is too far away from the *stationary* state (where production and dissipation almost match) which, according to common wisdom, is required in order to sustain spectral quasi-equilibrium. In fact, as shown more generally on SSVARTs in Sect. 9.3, this condition can be significantly relaxed, and furthermore, the relative mismatches in the RT and KH cases are similar (spectral quasi-equilibrium is seldom questioned for the latter).

[7] Ⓤ Despite numerous positive responses to introducing the "von Kármán number" terminology, the equally numerous negative opinions in the turbulence community have pushed us to use the "turbulence Knudsen number" in all further publications.

This ratio, which is presently known only from direct simulations for the RT case [33, 128–130], might have an error of up to 20%. The uncertainty on the K/K_D ratio in the RM case is at least of the same order [33], but it has a smaller effect on the von Kármán number.

As with the turbulent kinetic energy, which can be defined using either the one- or two-fluid formalism, the von Kármán number also has a two-fluid form, κ_b, obtained by replacing K with K_B in the expression of Λ_i in (4.10) (but not in (4.12), because the definition of E does not change). We then obtain the values given in Table 4.1, p. 38. This κ_b provides a more consistent reconstruction of the Knudsen number because, in certain models such as AWE's two-fluid, the directed kinetic energy is confined to large scales and thus is not associated with the inertial dissipation cascade.

The energy balances for the three instabilities (Table 4.1) give rise to several comments:

- In the Richtmyer–Meshkov case, K_D/K_I is time-dependent; it is necessary that $n_0 < {}^2\!/_3$ or else K_D will exceed K_I for $t \to \infty$, and thus K_D/K_I diverges for $t \to 0$. This shows that, unlike the KH and RT cases, the self-similar regime cannot be extended below a certain non-zero value of t; the system retains a memory of its initialization in the form of the input energy K_{I0}.

- Compared to the value in the the Kelvin–Helmholtz case, the ratio K_D/K_I is more than an order of magnitude greater in the Rayleigh–Taylor case $(4\mathcal{Y}_0/\mathcal{X}_0^2 \approx 50)$. Given the same input energy, the structure of velocity fluctuations, which include the turbulence and the directed interpenetration, is necessarily very different in the two situations. This element is crucial when comparing the one- and two-fluid models.

- As a consequence of the previous point, the one- and two-fluid von Kármán numbers are not noticeably different, except in the RT case.

- The von Kármán numbers in the KH and RM cases are of the same order as the von Kármán constant: about .4 [59, §5.2], [134, §17.1.4]. By contrast, in the RT case, the von Kármán number is smaller by about a factor of 5.[8] This agrees with qualitative observations on the respective sizes of eddies and bubbles in KH [110–117] and RT [33, 120–122] mixing zones.

---

[8] Ⓤ This fact, which does not seem to have been reported previously and is critical in our approach to model assessment, has stirred confusion among some readers. Confusion could stem from the apparent contradiction in bounding an inertial range with an integral length scale much smaller than the RT flow scale (the TMZ width): one could then be drawn to the unacceptable conclusion that the spectral density of turbulent energy is negligible in the wave number interval of large scales $[2\pi/L, 2\pi/\Lambda_i]$.

Actually there is a non negligible amount of energy in the large scales range, and this introduces an error on the integral length scale as calculated using (4.10). Assumptions on the reduced spectral density profile $e(k)$ (here k is the wave number) in the large scales range provide simple estimates of the possible deviations on the von Kármán number. The turbulent kinetic energy is given by:

4.4 "0D" Reduction of Evolution Equations of Hydrodynamic Models

The results for the "0D" mean values in Table 4.1, p. 38 make up a set of constraints with which any statistical model must comply. In general, matching and adjusting to these constraints, either by analytical or numerical means, can be physically complex [100–103]: the dynamics of a model is controlled by numerous competing effects (productions, dissipations, turbulent diffusions, advections), making the balance difficult to grasp, even for self-similar solutions of a simple model like the k–ε. However, the balances of Table 4.1 are "0D" approximations, so an exact verification is not absolutely necessary. Using a few simplifying assumptions, it is often possible to obtain evolution equations for L, K_I, and K_D by integrating the model equations over the space

$$K = \int_{k_L}^{k_i} C_k \varepsilon^{2/3} k_i^{-5/3} e(k) \mathrm{d}k + \int_{k_i}^{\infty} C_k \varepsilon^{2/3} k^{-5/3} \mathrm{d}k \ ,$$

where $k_L = A/L$, $k_i = A/\Lambda_i^*$, Λ_i^* defines the *actual* limit of the inertial range (as opposed to the *apparent* limit defined by Λ_i in (4.10)), C_k is the Kolmogorov constant, A is a constant adjusted to retrieve (4.10) for $e(k) = 0$, and $e(k)$ can be assumed to be either:

$$e(k) = \left|\begin{array}{ll} 0 & \text{Steep truncation of inertial spectrum,} \\ \dfrac{k - k_L}{k_i - k_L} & \text{Continuous transition with sub-}k_L \text{ modes (assumed negligible),} \\ 1 & \text{Flat large scales spectrum with steep truncation at } k_L. \end{array}\right.$$

Elementary calculation then provide the relationship between the actual and apparent von Kármán numbers, κ^* and κ respectively, for the three profiles of spectral density:

$$\kappa \approx \left|\begin{array}{l} \kappa^* \ , \\ \kappa^* \left(\dfrac{4 - \kappa^*}{3}\right)^{3/2} \ , \\ \kappa^* \left(\dfrac{5 - 2\kappa^*}{3}\right)^{3/2} \ , \end{array}\right.$$

using the approximation $\kappa^* \approx k_L/k_i$. These estimated corrections amount to less than 50% in the most extreme cases and always *reduce* the actual von Kármán number. The basic conclusion on the small value of the von Kármán number in RT flows is thus undisputable.

Incidentally, it is to be noticed that in the RT case, the sub-inertial range consists of *two* parts: the large scales range just discussed above, and the so-called frozen modes range $[0, 2\pi/L]$. This contrasts with practically all common turbulent shear driven flows, where the von Kármán number is close to one, and thus the large scales range is negligible and the sub-inertial range is reduced to frozen modes. For RT type flows, the large scales range is typically filled by the directed energy (hence $\kappa^* \approx \kappa_b$). As for any other turbulent flow, the dynamics of the system is dominated by the large scales motions, and therefore the specific presence of a large scales range demands to be taken into account in models. Two-fluid approaches answer this demand by separating the directed energy out of the turbulent energy.

coordinate. We have then the significant advantage of dealing with ordinary differential equations ("0D" model) rather than partial differential equations. This approach will be applied to DAM's k–ε and AWE's two-fluid models, respectively, in Sects. 5.2 and 6.1.

To illustrate the procedure of 1D to "0D" projection, let us consider the example of a generic per volume quantity a that is described in a given model by an evolution equation having source and flux terms:

$$\tfrac{\partial}{\partial t} a + (\phi_x^a)_{,x} = s^a \ . \tag{4.13}$$

By integrating over the variable interval $[-L(t)/2, +L(t)/2]$, the evolution equation leads to:

$$\frac{1}{L} \int_{-L/2}^{+L/2} \tfrac{\partial}{\partial t} a(x)\,\mathrm{d}x + \frac{\phi_x^a(+L/2) - \phi_x^a(-L/2)}{L} = \frac{1}{L} \int_{-L/2}^{+L/2} s^a(x)\,\mathrm{d}x \ , \tag{4.14}$$

It is assumed that the fluids are at rest outside of the mixing zones for the three flows under consideration. In these zones of rest, we assume that the quantities of interest (a, in this case) cancel one another and yield flux and source terms of zero. This eliminates the flux term and allows us to rearrange the integral and the time derivative. Introducing the averaged values for the zone, $A = \langle a \rangle$ and $S^a = \langle s^a \rangle$, there appears a term in $\mathrm{d}L/\mathrm{d}t$:

$$\frac{1}{L} \frac{\mathrm{d}(LA)}{\mathrm{d}t} = \frac{\mathrm{d}A}{\mathrm{d}t} + \frac{\mathrm{d}L}{L\,\mathrm{d}t} A = S^a \ . \tag{4.15}$$

Thus, comparing with the evolution equation, we see that the flux term is replaced by a mean "dilution" term in $(\mathrm{d}L)/(L\mathrm{d}t)$. This last term is then closed by the expression for $L'(t)$ in (4.5).[9]

The difficulty now is to transpose the closure of s^a to S^a, which must be expressed realistically in terms of other mean quantities. One systematic way of proceeding is to consider the mean values as equal to the values at *the center of the mixing zone* to within a constant factor ζ. This is equivalent to stating that there is some dominant mode which defines the generic profile of any quantity, as was implicitly assumed in [100] and [101–103]. Assuming that s^a is a function $s^a(b)$ of another quantity b, we then have:

$$S^a \approx \frac{1}{\zeta} s^a\left(\zeta\,B\right) \ . \tag{4.16}$$

---

[9] It might seem that the modeling of the flux terms of the various quantities is lost in the "0D" equation (4.15). Actually, the closure of $L'(t)$ in (4.5) links all the fluxes to the velocity δU. The "0D" reduction thus introduces distortions when the fluxes of the various quantities are not consistent with a single mean advection velocity; this is the case for instance with the k–ε model applied to a Kelvin–Helmholtz instability (see Sect. 5.2) for which corrections can be introduced (see C).

It must therefore be emphasized that only the source terms are approximated in the 1D→0D projection procedure. As most of the values are bell-shaped and cancel at the edges of the mixing zone, the simplest profile is a parabola $\alpha^+(x)\alpha^-(x)$, and we have $\zeta \approx 3/2$. This value will be used throughout this work.

5

Reconstruction by DAM's k–ε Model of Developed Mixing Instabilities

5.1 On the "Richtmyer–Meshkov" Calibration of Coefficients $C_{\varepsilon 0}$ and σ_ϱ

According to the literature, the calibration of coefficients of the incompressible k–ε model (listed in Table 2.1, p. 14) is based on the reconstruction of a *set of canonical experiments representative of the set of phenomena to be captured, excluding all other effects* that are not described or are possibly spurious. The content of this set of experiments varies among authors, but there are three main classes:

- experiments sensitive to just one simple phenomenon, most often self-similar, observed over a long period and reconstructed from a single parameter that is a function of the model coefficients (decay of grid generated turbulence, logarithmic zone of the turbulent boundary layer, homogeneous shear of a turbulent flow, etc.);
- experiments sensitive to several phenomena, usually reconstructed by numerical optimization, simple parameters being predetermined (accelerated boundary layers, etc.);
- "gedanken" experiments sensitive to the internal consistency of the model and reconstructed theoretically (diffusion–dissipation competition between k and ε, diffusion of a passive tracer in isotropic homogeneous turbulence by the renormalization group, etc.).

Quite naturally, the strategy for calibrating DAM's k–ε model thus consisted in complementing the set of canonical k–ε incompressible calibrations with experiments sensitive to the coefficients $C_{\varepsilon 0}$ and σ_ϱ.[1] Richtmyer–Meshkov mixing instabilities observed in shock tubes have been considered of central

---

[1] The coefficient $C_{\varepsilon 3}$ has little influence in the instabilities under consideration and can be fairly well estimated by a "gedanken" experiment, as will be elaborated in Sect. 8.4.

A. Llor: *Statistical Hydrodynamic Models for Developed Mixing Instability Flows*,
Lect. Notes Phys. **681**, 47–61 (2005)
www.springerlink.com

importance because they resemble phenomena that occur in many DAM applications. However, in addition to the experimental difficulties [135] (membrane influence, boundary layers, observation time, diagnostics, departure from self-similarity, etc.), these experiments have several disadvantages:

- the self-similar behavior observed after shock passage does not depend on the coefficients $C_{\varepsilon 0}$ and σ_{ϱ}, as will be seen in Sect. 5.2 ("RM" column of Table 5.1, p. 52);
- the turbulence production by the shock, which is in fact the phenomenon to be reconstructed by the calibration, is tainted by several difficulties associated with the relevance of the modeling of the so-called "Rayleigh–Taylor" term, as will be analyzed in Sect. 7.2;
- assuming that the two preceding items are amendable, the turbulence production by shock depends on *both* coefficients $C_{\varepsilon 0}$ and σ_{ϱ}, whereas so far calibrations have been based on the ensuing evolution of the mixing zone width *only*. Additional intricate measurements [135–138] (in particular, velocity fluctuations) or an additional independent experiment are necessary in order to define the coefficients unambiguously.

To remove the ambiguity in the calibration of the coefficients $(\sigma_{\varrho}, C_{\varepsilon 0})$ by shock-tube experiments, some authors have considered other types of flows. For instance, the decay of stratified grid turbulence provides the empirical relationship [5, 6]:

$$C_{\varepsilon 0} = 1 + \sigma_{\varrho}/5 , \tag{5.1}$$

but it assumes that the so-called "Rayleigh–Taylor" production term remains active when $(\overline{p})_{,i}(\overline{\varrho})_{,i} > 0$ (which contradicts item 11 in Sect. 2.3), and it implies that $C_{\varepsilon 0} > 1$. The presently recommended value $C_{\varepsilon 0} = .85$ was proposed by C.E. Leith for reconstructing $L(t)$ for a self-similar Rayleigh–Taylor instability, but with certain assumptions about some coefficients, such as $\sigma_{\varrho} = \sigma_c = .7$ and $C_{\varepsilon 1} = 2$ [12]. Other authors [100] have studied the Rayleigh–Taylor self-similar instability in shock tube experiments, but obtained different values.

In fact the Rayleigh–Taylor self-similar instability does not have any of the above-mentioned disadvantages of the Richtmyer–Meshkov instability. As will be seen in Sect. 5.3, this makes it possible to calibrate both $C_{\varepsilon 0}$ and σ_{ϱ} by simultaneous reconstruction of the growth of $L(t)$ and of another parameter, energy balance or von Kármán number (integral length scale). The complexity of the 1D calculations suggests the use of simplifications or numerical simulations [100–103], and we will limit ourselves to illustrating this calibration method in the "0D" approximation.

5.2 "0D" Reconstruction by DAM's k-ε Model of Basic Self-Similar Instabilities

5.2.1 General Remarks, Reconstruction of the Directed Kinetic Energy

The k-ε model described in Sects. 2.2 and 2.3 comprises six evolution equations, of which only those in $\tilde{k}$ and $\tilde{\varepsilon}$ remain relevant after "0D" projection according to the procedure described in Sect. 4.4. Indeed:

- forcing $\alpha^{\pm}(x)$ and $U_x^{\pm}(x)$ to follow some given profiles is equivalent to solving the concentration and mean momentum equations along x, and
- assuming incompressibility makes the internal energy equation useless, because it is merely a passive receptacle for dissipative phenomena.

The remaining mean quantities, $K = \langle \tilde{k} \rangle$ and $E = \langle \tilde{\varepsilon} \rangle$ are not directly comparable to the mean quantities in Table 4.1, p. 38, K_I and K_D, which were deduced from the self-similar growth of the mixing zone. Thus we reconstruct K_I and K_D from other parameters of the model.

The difference between fluid velocities is related to the turbulent concentration flux, which determines the growth of the mixing zone. This relation will be detailed and extended in (7.6), Sect. 7.3. For now, we write at the center of the mixing zone:

$$U_x^+ - U_x^- = \overline{\frac{\varrho c^+ u_x}{\varrho c^+}} - \overline{\frac{\varrho c^- u_x}{\varrho c^-}} = \overline{\frac{\varrho c^+ u_x''}{\varrho c^+}} - \overline{\frac{\varrho c^- u_x''}{\varrho c^-}}$$

$$\overset{\mathrm{m}}{=} -\frac{C_\mu}{\sigma_c} \frac{\tilde{k}^2}{\tilde{\varepsilon}} \left[\frac{(C^+)_{,x}}{C^+} - \frac{(C^-)_{,x}}{C^-} \right]$$

$$\overset{\mathrm{m}}{=} -4\zeta \frac{C_\mu}{\sigma_c} \frac{K^2}{E L}, \tag{5.2}$$

the last two expressions being obtained by transforming $\overline{\varrho c^{\pm} u_x}$ with the Boussinesq–Reynolds closure in (2.16) and the "0D" approximation procedure described in Sect. 4.4. The growth of the mixing zone thickness, $L(t)$, and the directed kinetic energy, K_D, are then deduced from their respective expressions (4.5) and (4.6a).

5.2.2 Construction and Resolution of the "0D" Differential System

According to the expressions in Table 4.1, p. 38 that are valid in the "0D" approximation, the input energy, K_I, is directly determined by the mixing zone thickness. In the k-ε model, this must be made identical to the time integral of the turbulent energy production term, Π_K. With an accuracy equal to the "0D" approximations, this is rigorous for the Kelvin–Helmholtz case, because Π_K is the work of the Reynolds stresses, which converts mean kinetic energy

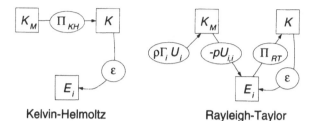

Fig. 5.1. Diagram of energy transfers in Kelvin–Helmholtz and Rayleigh–Taylor instabilities according to DAM's k-ε model. In the limit of zero Atwood number and incompressible fluids, the RT case should have $\Pi_K = \varrho\Gamma U$, or $K_I^* = K_I$

into turbulent kinetic energy. For the Rayleigh–Taylor case, the energy transfers modeled in DAM's k-ε model are more complicated: as shown in Fig. 5.1 and discussed in Sect. 7.2, the energy is first input into mean kinetic energy by the work of gravity, then transferred into internal energy by Favre averaged compression $(U_i)_{,i}$, and finally converted into turbulent kinetic energy by the so-called "Rayleigh–Taylor" production term. In the case of incompressible fluids, these transfers must all be identical so that the work of the gravity forces is entirely converted into K (the dissipation $\tilde{\varepsilon}$ is the only source of internal energy, and the mean kinetic energy K_M is asymptotically zero for $\mathcal{A} \to 0$). Because the k-ε model does not impose this identity a priori, it must therefore be introduced by a suitable choice of coefficients, which will be discussed in Sect. 5.3. The time integral of Π_K will be denoted K_I^* to distinguish it from K_I.

Still assuming that the mixing zone displays parabolic profiles for $\tilde{k}$ and $\tilde{\varepsilon}$ and linear profiles for $\alpha^{\pm}$ as in Fig. 4.1, we obtain in each case the expressions for the turbulent energy source term, Π_K. For the Kelvin–Helmholtz instability, we thus obtain the term for mean specific production by Reynolds stresses from (2.22):

$$\Pi_K^{\boxed{KH}} \underset{\equiv}{} \frac{1}{\zeta}\left[C_\mu \frac{\tilde{k}^2}{\tilde{\varepsilon}}\left[(U_y)_{,x}\right]^2\right]_{x=0} \approx C_\mu \frac{K^2}{E}\left(\frac{\Delta U_y}{L}\right)^2 . \qquad (5.3)$$

For the Rayleigh–Taylor instability, the gradients of $\bar{p}$ and $\bar{\varrho}$ are uniform and given for $\mathcal{A} \to 0$ by:[2]

---

[2] The expression $P(+L/2) - P(-L/2)$ is obtained by integrating the momentum equation in (2.6) over the mixing zone thickness:

$$\frac{1}{L}\int_{-L/2}^{+L/2}\left[\tfrac{\partial}{\partial t}\bar{\varrho}U_x + (\bar{\varrho}U_xU_j)_{,j} + (\bar{p})_{,x} + \overline{(\varrho u_x''u_j'')}_{,j} - \bar{\varrho}g_x\right]\mathrm{d}x = 0 .$$

However, in $\pm L/2$, $U_x = 0$ and $\overline{\varrho u_x''u_j''} = 0$, and with the linear approximation in Fig. 4.1:

$$\left[\frac{(\overline{\varrho}),_x}{\overline{\varrho}}\right]_{x=0} \approx \frac{(\varrho^+ - \varrho^-)/L}{(\varrho^+ + \varrho^-)/2} = \frac{2\mathcal{A}}{L},$$

$$\left[\frac{(\overline{p}),_x}{\overline{\varrho}}\right]_{x=0} \approx \frac{[P(+L/2) - P(-L/2)]/L}{(\varrho^+ + \varrho^-)/2} \approx -\Gamma, \tag{5.4}$$

from which the mean per mass production term deduced from (2.17) is:

$$\Pi_K^{\boxed{RT}} \stackrel{m}{=} -\frac{1}{\zeta}\left[\frac{C_\mu}{\sigma_\varrho} \frac{\tilde{k}^2}{\tilde{\varepsilon}} \frac{(\overline{p}),_x (\overline{\varrho}),_x}{\overline{\varrho}^2}\right]_{x=0} \approx \frac{2C_\mu}{\sigma_\varrho} \frac{K^2}{EL} \mathcal{A}\Gamma. \tag{5.5}$$

Furthermore, it can be shown that in the Rayleigh–Taylor case, the production by *modeled* Reynolds stresses is a negligible contribution compared with Π_{RT} [5,6].[3] For the Richtmyer–Meshkov instability, $\Pi_K = 0$, because there is production only when the initial shock passes, mixing taking place afterwards under the sole effect of relaxing turbulence. An analysis of this initial production, which does not involve the self-similar approach, will be given in Chap. 8.

Finally, following the general procedure described in Sect. 4.4, the "0D" version of the k–ε model reduces to the differential system:

$$\begin{cases} \dfrac{d}{dt}L = \dfrac{8\zeta C_\mu}{\sigma_c} \dfrac{K^2}{EL}, \\[2mm] \dfrac{d}{dt}K_I^* = -\dfrac{dL}{L\,dt}K_I^* + \Pi_K, \\[2mm] \dfrac{d}{dt}K = -\dfrac{dL}{L\,dt}K + \Pi_K - E, \\[2mm] \dfrac{d}{dt}E = -\dfrac{dL}{L\,dt}E + C_{\varepsilon\pi}\dfrac{E}{K}\Pi_K - C_{\varepsilon 2}\dfrac{E^2}{K}. \end{cases} \tag{5.6}$$

For each type of self-similar mixing flow, we only need insert into this differential system the expressions and models of Π_K and $C_{\varepsilon\pi}$, as well as the appropriate power laws in t for L, K_I^*, K, and E. The results of this substitution, which is lengthy and tedious but straightforward, are shown in Table 5.1, along with the von Kármán numbers as defined by (4.11) to characterize the turbulence structure. Various significant numerical values corresponding to the set of coefficients in Table 2.1, p. 14 are also given.

$$\frac{1}{L}\int_{-L/2}^{+L/2} \overline{\varrho}\, U_x\, dx \approx -\frac{\mathcal{Y}\mathcal{A}^2}{3}\frac{\varrho^+ + \varrho^-}{2}\Gamma t,$$

which is of second order in $\mathcal{A}$. This gives the approximate expression of $(\overline{p}),_i$ that is valid in the limit $\mathcal{A} \to 0$.

[3] The estimation of the mean momentum in footnote 2, p. 50 shows that in the Rayleigh–Taylor case, the production term due to Reynolds stresses is of order $\mathcal{A}^4$. In the limit $\mathcal{A} \to 0$, it is thus negligible compared with the so-called "Rayleigh–Taylor" term, which is of order $\mathcal{A}^2$.

Table 5.1. Basic geometric and energy characteristics of KH, RT, and RM turbulent mixing zones as reconstructed by the "0D" simplified k–ε model. The boxed numerical values were obtained with the standard coefficients of the k–ε model in Table 2.1, p. 14. These results are to be compared with the "0D" experimental data in Table 4.1, p. 38

Instability	KH	RT	RM
Constant	$\mathcal{X}_{k\varepsilon} = \sqrt{\dfrac{C_\mu}{C_{\varepsilon 1} C_{\varepsilon 2}}(C_{\varepsilon 2} - C_{\varepsilon 1})}$ $\boxed{\approx .077}$	$\mathcal{Y}_{k\varepsilon} = \dfrac{2C_\mu}{\sigma_\varrho}\dfrac{(C_{\varepsilon 2} - C_{\varepsilon 0})^2}{(4C_{\varepsilon 2} - 3)(4C_{\varepsilon 0} - 3)}$ $\boxed{\approx .054}$	$n_{k\varepsilon} = \dfrac{2C_{\varepsilon 2} - 3}{3C_{\varepsilon 2} - 3}$ $\boxed{\approx .30}$
L	$\mathcal{X}_{k\varepsilon}\,\Delta U_y t$	$\mathcal{Y}_{k\varepsilon}\, A\Gamma t^2$	$L_0\left(\dfrac{t}{t_0}\right)^{n_{k\varepsilon}}$
K_I^*	$[\sigma_c] \times \dfrac{1}{12} \times (\Delta U_y)^2$ $\boxed{\sigma_c \approx .7}$	$\left[\dfrac{\sigma_c}{\sigma_\varrho}\right] \times \dfrac{\mathcal{Y}_{k\varepsilon}}{12} \times (A\Gamma t)^2$ $\boxed{\sigma_c/\sigma_\varrho \approx .35}$	$K_{I0}\left(\dfrac{t}{t_0}\right)^{-n_{k\varepsilon}}$
K_D	$\dfrac{\mathcal{X}_{k\varepsilon}^2}{48} \times (\Delta U_y)^2$	$\dfrac{\mathcal{Y}_{k\varepsilon}^2}{12} \times (A\Gamma t)^2$	$\dfrac{n_{k\varepsilon}^2}{48} \times \left(\dfrac{L_0}{t_0}\right)^2\left(\dfrac{t}{t_0}\right)^{-2(1-n_{k\varepsilon})}$

Table 5.1. continued

Instability	KH		RT		RM	
K/K_J^*	$\dfrac{C_{\varepsilon 2} - C_{\varepsilon 1}}{C_{\varepsilon 2}}$	$\approx .226$	$\dfrac{4(C_{\varepsilon 2} - C_{\varepsilon 0})}{4C_{\varepsilon 2} - 3}$	$\approx .913$		
K/K_D	$\dfrac{4\sigma_c}{C_\mu} \dfrac{C_{\varepsilon 1}}{C_{\varepsilon 2} - C_{\varepsilon 1}}$	≈ 106	$\dfrac{2\sigma_c}{C_\mu} \dfrac{4C_{\varepsilon 0} - 3}{C_{\varepsilon 2} - C_{\varepsilon 0}}$	≈ 5.93	$\dfrac{12\sigma_c}{C_\mu} \dfrac{1}{2C_{\varepsilon 2} - 3}$	≈ 117
κ	$\sqrt{\dfrac{\sigma_c}{12C_\mu} \dfrac{C_{\varepsilon 2} - C_{\varepsilon 1}}{C_{\varepsilon 1}}}$	$\approx .44$	$\sqrt{\dfrac{\sigma_c}{6C_\mu} \dfrac{C_{\varepsilon 2} - C_{\varepsilon 0}}{4C_{\varepsilon 0} - 3}}$	≈ 1.84	$\sqrt{\dfrac{\sigma_c}{36C_\mu} (2C_{\varepsilon 2} - 3)}$	$\approx .42$

5.2.3 Realizability of Turbulent Closures

We can now return to the realizability of the Boussinesq–Reynolds closure of the various turbulent fluxes and the possible limitations that may be required in the "0D" model. For example, for the turbulent concentration fluxes $\overline{\varrho c^{\pm} u_i''}$, realizability requires that at the center of the mixing zone (where $C^{\pm}$ gradients are strongest):

$$\left| \overline{\varrho c^{\pm} u_i''} \right| \stackrel{\mathrm{m}}{=} \frac{C_\mu}{\sigma_c} \, \overline{\varrho} \, \frac{\tilde{k}^2}{\tilde{\varepsilon}} \, |(C^{\pm})_{,x}| < \overline{\varrho} \sqrt{\tilde{k}} \, C^{\pm} . \tag{5.7}$$

Introducing the "0D" mean values, we find that for all types of mixing zones:

$$\kappa < \frac{\sigma_c}{2\sqrt{\zeta}\,C_\mu} \approx 3.17 . \tag{5.8}$$

In the KH and RM cases, the von Kármán number values of $\kappa \approx .44$ and $.42$ ensure realizability of practically all turbulent fluxes. Furthermore, in the KH case, the mean Reynolds tensor can be written using the results in Table 5.1 as:

$$R_{ij} = \frac{\varrho^+ + \varrho^-}{3} \, K \times \begin{pmatrix} 1 & \frac{3}{2}\sqrt{\frac{C_{\varepsilon 2}C_\mu}{C_{\varepsilon 1}}} \\ \frac{3}{2}\sqrt{\frac{C_{\varepsilon 2}C_\mu}{C_{\varepsilon 1}}} & 1 \end{pmatrix} . \tag{5.9}$$

The eigenvalues of R_{ij}:

$$R_{\pm} = \frac{\varrho^+ + \varrho^-}{3} \, K \times \left(1 \pm \frac{3}{2}\sqrt{\frac{C_{\varepsilon 2}C_\mu}{C_{\varepsilon 1}}} \right) \approx \frac{\varrho^+ + \varrho^-}{3} \, K \times (\, 1.51 \text{ or } .49 \,) , \tag{5.10}$$

are thus positive (therefore, R_{ij} is realizable) despite an appreciable anisotropy ratio, of the order of 3, which agrees with experimental data [110–117]. In the RT case, the von Kármán number ($\kappa \approx 1.84$, which is greater than 1 for the coefficients recommended in Table 2.1, p. 14) also ensures the realizability of the turbulent concentration flux, but with a small safety factor. Besides, regarding the closure of the turbulent mass flux involved in Π_K in (5.4), we note the presence of the Atwood number, which always ensures realizability in the limit $\mathcal{A} \to 0$. Thus the study presented here is not constrained by the realizability of closures for any type of instability.

5.2.4 Estimation of Errors Introduced by the "0D" Approach

It is now important to estimate the biases and uncertainties introduced by the "0D" projection. For this we rely on a comparison with the experimental findings in the Kelvin–Helmholtz case: indeed, the contribution of the model to the observed distortions in this case is minimal, since the k-ε model was developed and calibrated primarily to reconstruct turbulent boundary layers, which are similar to sheared mixing layers. A comparison of the "KH" columns in Tables 4.1, p. 38 and 5.1 yields the following results:

- the mixing zone growth rate, given by $\mathcal{X}_{k\varepsilon}$, is underestimated by about 23%,
- the balance of total energy input into the mixing zone, given by the constant σ_c, is underestimated by about 30%,
- the von Kármán number is underestimated by about 30%.

The agreement is generally satisfactory, considering the large approximations introduced in the "0D" projection. The underestimated growth of the mixing zone might be attributed to the experimental uncertainties, especially those related to the establishment of a developed self-similar regime [110–117].

Regarding the underestimation of the energy K_I^* with respect to K_I, it must be noted that the balance is restored by imposing $\sigma_c = 1$, which is too high and unacceptable considering the precision of common calibrations. However, the estimate of K_I given by (4.8) assumes that the gradients of *all* the relevant values (concentrations, velocities, etc.) are identical, as explained in Sect. 4.2. Actually, an elementary turbulence model such as k–ε introduces differences between profiles, as shown in C. Thus there is a certain ambiguity in the "0D" expressions for the mixing zone thickness and the input energy associated with it. This effect depends on the type of instability being considered. In the Kelvin Helmholtz case, the natural definition of L must be associated with the *transverse velocity profile*, $U_y(x)$, as was actually realized in the analysis of the experimental results [110–117]. Because the Schmidt–Prandtl coefficient of the momentum is defined as equal to 1, we can assume that $\sigma_c = 1$ in all the "0D" reasoning above. The energy balance in the KH case is then correctly captured by the "0D" k–ε model, with a von Kármán number of $.44/\sqrt{.7} \approx .53$, which is only 15% lower than the experimental value.

5.3 "0D Rayleigh–Taylor" Calibration of Coefficients $C_{\varepsilon 0}$ and σ_ϱ

5.3.1 Discussion of the "0D" Results

The expression for the "0D" Youngs constant, $\mathcal{Y}_{k\varepsilon}$ in Table 5.1, p. 52, was previously obtained using a simplified 1D approach (where $\sigma_k = \sigma_\varepsilon$) and confirmed by comparison with 1D numerical simulations [100]. Similarly, the self-similar exponent of the RM instability was found to be identical to within 3% to that obtained by 1D analytic calculations [101–105]. For non-negligible Atwood numbers, another 1D study yielded implicit analytic expressions which coincide with those of Table 5.1, p. 52 for the RT and RM cases when $\mathcal{A} \to 0$ [106, 107]. This confirms the elements presented above in Sect. 5.2 regarding the minor distortions seen in the "0D" approach. In contrast to the KH case, which required the substitution $\sigma_c = 1$, σ_c retains its usual value for the RT case, because the profile of $C'^{\pm}$, and not that of U_y, is now appropriate for calculating the potential energy balance.

The coefficients usually recommended in DAM's k–ε model underestimate the mixing zone thickness $L(t)$ and the input energy K_I^* by factors of about

2 and 3, respectively (see Table 5.1, p. 52). These values are barely compatible with experimental and "0D"-projection uncertainties, and one would have reasonably expected a better reconstruction from a model that was specially designed to capture this kind of instability. It must be emphasized that the factor of 3 on energy is not associated with the smaller growth rate of the mixing zone (already taken into account by the factor $\mathcal{Y}_{k\varepsilon}$ in Table 5.1, p. 52), and actually represents a *deficit* of input energy compared with the potential energy balance deduced from the concentration profile (see Sect. 5.2). Regarding the von Kármán number, the k–ε model overestimates it by a factor of 14, which is far above the uncertainties for this quantity obtained from direct numerical simulations [33]. This reflects the high K/K_I^* ratio and the low dissipation of turbulent energy.

One may ask whether a better set of coefficients $C_{\varepsilon 0}$ and σ_ϱ might suitably reconstruct the principal quantities. In Fig. 5.2, the combinations that verify various constraints are shown in the $(C_{\varepsilon 0}, \sigma_\varrho)$ plane. We note that it is impossible to reconstruct simultaneously the mixing zone growth, the input energy balance (determined by σ_ϱ), and the von Kármán number (determined by $C_{\varepsilon 0}$). Various $(C_{\varepsilon 0}, \sigma_\varrho)$ combinations can thus be used depending on how are prioritized the quantities to be captured.

5.3.2 Stability and the RT Case with Self-Similar Variable Acceleration[4]

Another important consideration in the choice of $(C_{\varepsilon 0}, \sigma_\varrho)$ is model stability. This problem is difficult to analyze in general, but becomes simpler for self-similar incompressible solutions in the "0D" projection. In order to somewhat represent the different flow forcings encountered in applications, such as time-dependent accelerations, we shall thus look for solutions of system (5.6) under self-similar evolutions of Γ as t^n. Substituting into (5.6) the corresponding behaviors in $K \propto t^{2n+2}$, $E \propto t^{2n+1}$, and $L \propto t^{n+2}$, lengthy and tedious though straightforward calculations yield:

$$
\begin{cases}
L = \dfrac{2C_\mu}{\sigma_\varrho} \dfrac{(C_{\varepsilon 2} - C_{\varepsilon 0})^2}{[(3C_{\varepsilon 2}-3)n+(4C_{\varepsilon 2}-3)][(3C_{\varepsilon 0}-3)n+(4C_{\varepsilon 0}-3)]} \, A\Gamma t^2 = \mathcal{Y}_n A\Gamma t^2, \\[2ex]
K_I^* = \left[\dfrac{\sigma_c}{\sigma_\varrho}\right] \times K_I, \\[2ex]
K = \left[\dfrac{\sigma_c}{\sigma_\varrho}\right] \dfrac{(3n+4)(C_{\varepsilon 2} - C_{\varepsilon 0})}{(3C_{\varepsilon 2}-3)n+(4C_{\varepsilon 2}-3)} \times K_I, \\[2ex]
\kappa = \sqrt{\dfrac{\sigma_c}{12C_\mu} \dfrac{(n+2)(C_{\varepsilon 2} - C_{\varepsilon 0})}{(3C_{\varepsilon 0}-3)n+(4C_{\varepsilon 0}-3)}}.
\end{cases}
$$

$$(5.11)$$

---

[4] Ⓤ New developments on SSVARTs [54,55] discussed in Sect. 9.3 have noticeably expanded the findings of this part.

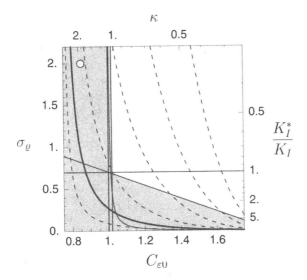

Fig. 5.2. Graphical representation in the $(C_{\varepsilon 0}, \sigma_\varrho)$ plane of the different constraints stemming from the "0D" analysis of the self-similar Rayleigh–Taylor instability by DAM's k–ε model: (*empty circle*) recommended coefficient values; (*heavy line*) $(C_{\varepsilon 0}, \sigma_\varrho)$ values that reconstruct the observed mixing zone growth rate, at $\mathcal{Y}_{k\varepsilon} = .12$; (*dashed lines*) $(C_{\varepsilon 0}, \sigma_\varrho)$ values that reconstruct the observed mixing zone growth rate to within a constant factor of: 3, .3, .1, .03, and .01, respectively, from left to right; (*grayed area*) $(C_{\varepsilon 0}, \sigma_\varrho)$ values leading to, *for self-similar instabilities with acceleration variable as t^n,* unphysical quantities (negative turbulent energy), non-realizable turbulent fluxes, violation of the second law of thermodynamics, or mixing zone growth faster than free fall. The von Kármán number and the energy balance are associated unambiguously with $C_{\varepsilon 0}$ and σ_ϱ, respectively, and their scales are thus placed oppositely. (*horizontal line*) at $\sigma_\varrho = \sigma_c \approx .7$ corresponds to the correct energy balance, $K_I^* / K_I = 1$

We have introduced here constant $\mathcal{Y}_n$, which is characteristic of the mixing zone growth rate, and expressed the energies by means of the total input energy, K_I, which now reads:

$$K_I = \frac{-2}{(\varrho^+ + \varrho^-)L} \int_0^t \Gamma(s) \int_{-\mathcal{Y}_n A\Gamma(t)t^2/2}^{+\mathcal{Y}_n A\Gamma(t)t^2/2} \bar{\varrho}(s,x) U(s,x) \, \mathrm{d}x \, \mathrm{d}s$$

$$= \frac{(n+2)\mathcal{Y}_n}{6(3n+4)} \left[A\Gamma(t)t \right]^2 . \tag{5.12}$$

Compared with the ordinary Rayleigh–Taylor case, the formulas in (5.11) yield equivalent results for the input energy K_I^*. However, the presence of the factor $[(3C_{\varepsilon 0} - 3)n + (4C_{\varepsilon 0} - 3)]$ in $\mathcal{Y}_n$ and κ, which are respectively positive and real quantities, requires that:

$$\forall n \geq 0, \; C_{\varepsilon 0} > \frac{3n+3}{3n+4} \implies \boxed{C_{\varepsilon 0} \geq 1,} \tag{5.13}$$

instead of the previous $C_{\varepsilon 0} > {}^{3}/{4}$ for $n = 0$. The recommended value of $C_{\varepsilon 0} = .85$ results in positive solutions provided that $n < -(4C_{\varepsilon 0}-3)/(3C_{\varepsilon 0}-3) \approx .89.$[5]

The constraint (5.13) can be tightened a little by noting that $C_{\varepsilon 0}$, being related to the von Kármán number, controls the realizability of first gradient closures of turbulent fluxes by the condition (5.8). Applying (5.8) to the expression for κ in (5.11) yields:

$$\forall n \geq 0, \; \frac{\sigma_c}{12C_\mu} \frac{(n+2)(C_{\varepsilon 2} - C_{\varepsilon 0})}{(3C_{\varepsilon 0}-3)n+(4C_{\varepsilon 0}-3)} < \frac{\sigma_c^2}{4\zeta C_\mu^2}$$

$$\implies \boxed{C_{\varepsilon 0} \geq \frac{9\sigma_c + \zeta C_\mu C_{\varepsilon 2}}{9\sigma_c + \zeta C_\mu}.} \tag{5.14}$$

---

[5] It must be noted that the nonpositivity of turbulent quantities for self-similar solutions does not mean that a nonpositivity actually appears in model calculations. A case initialized with positive values of K and E retains this property because of the homogeneous nature of the k–ε equations, but a stability study then shows that the solution of the equations does not converge towards a self-similar evolution: there is, for example, an asymptotic decay of E toward 0. Naturally, this type of evolution is as non-physical as is the nonpositivity of the self-similar solution. A linear stability study was made in the vicinity of the self-similar solution. This study, tedious but with no conceptual difficulties, will not be presented here, because it does not provide any additional constraint beyond positivity as given in (5.13).

Nevertheless, one might find debatable that a model such as k–ε should necessarily remain stable for all self-similar external forcings which, in the absence of experimental data, may display instabilities in a real flow. However, the k–ε is a statistical model which must remain stable, because it describes the *ensemble mean of all realizations* of a flow (even if, as such, the model may be irrelevant to particular flow realizations, when involving symmetry breaking for instance): the mean of chaos is not chaotic!

Finally, the constraint according to which the coefficients C_ε associated with production terms must be greater than 1 is a universal fact, to be attributed to the behavior of the integral time K/E. Under the sole influence of a single production term in the equations of K and E and in the absence of any other characteristic time scale, the turbulent relaxation time *increases* with the turbulent energy if $C_\varepsilon < 1$; this contradicts the general principles of physical theories (classical Lagrangian mechanics, electromagnetism, quantum mechanics, relativity, etc.), which always yield an inverse time–energy relationship.

In the case of an RT flow accelerating as t^n, it must be also noted that the characteristic time of the instability increases and that the "Rayleigh–Taylor production" term dominates in the limit $n \to \infty$: thus we see that it is not acceptable to have $C_{\varepsilon 0} < 1$, which would make the integral time decrease.

Regarding the integral length scale, $K^{3/2}/E$, one can imagine that it cannot be noticeably modified by the production term through Reynolds stresses, and consequently, $C_{\varepsilon 1} \approx {}^{3}/{2}$. Calibrations at $C_{\varepsilon 0} \approx 1$ thus tend to increase the integral length scale. At this level there is therefore a fundamental difference between the driving terms of the Kelvin–Helmholtz and Rayleigh–Taylor instabilities of DAM's k–ε model.

One can imagine that limiting the turbulent concentration flux in a case that violates the realizability constraint (5.8) would make it possible to keep a suitable reconstruction of the instability and to relax (5.14). However, this situation would be unacceptable, because limiting the turbulent concentration flux eliminates the constant σ_c and thus completely changes the relationship between K_I and K_I^* in (5.11).[6]

Another stability condition, less direct but of general physical bearing, is to satisfy the second law of thermodynamics. In the present case, with the incompressibility assumption, the second law translates into a *non decreasing internal energy*. Referring back to the energy flux diagram in Fig. 5.1, this leads simply to the condition $K < K_I$, which from (5.11) is expressed as:

$$\forall n \geq 0, \ \sigma_\varrho > \sigma_c \frac{(3n+4)(C_{\varepsilon 2}-C_{\varepsilon 0})}{(3C_{\varepsilon 2}-3)n+(4C_{\varepsilon 2}-3)} \implies \boxed{\sigma_\varrho \geq \sigma_c \frac{C_{\varepsilon 2}-C_{\varepsilon 0}}{C_{\varepsilon 2}-1}}. \quad (5.15)$$

Although not strictly speaking a stability constraint, it must also be required that the mixing zone growth remains slower than free fall. From the expression for $\mathcal{Y}_n$ in (5.11), this leads to:[7]

$$\forall n \geq 0, \ \mathcal{Y}_n < \frac{2}{(n+1)(n+2)} \implies \boxed{\sigma_\varrho \geq \frac{C_\mu}{9} \frac{(C_{\varepsilon 2}-C_{\varepsilon 0})^2}{(C_{\varepsilon 2}-1)(C_{\varepsilon 0}-1)}}. \quad (5.16)$$

The conditions of positivity, realizability of turbulent fluxes, compatibility with the second law of thermodynamics, and growth slower than free fall, (5.13), (5.14), (5.15), and (5.16), are shown in Fig. 5.2. One can thus see that no pair of coefficients $(C_{\varepsilon 0}, \sigma_\varrho)$ can reconstruct the growth of the Rayleigh–Taylor mixing zone. The observed distance between the line $\mathcal{Y}_{k\varepsilon} = .12$ and

---

[6] A detailed calculation where the equation in L of (5.6) is replaced by:

$$\frac{\mathrm{d}}{\mathrm{d}t} L = 2\sqrt{K} \ ,$$

stemming from the limitation of $\overline{\varrho c^{\pm} u_i''}$ in (5.7), yields a growth rate $\mathcal{Y}_n$ identical to that in (5.11), but changes K_I^* and κ into:

$$\begin{cases} K_I^* = \dfrac{C_\mu}{\sigma_\varrho} \dfrac{3(n+2)(C_{\varepsilon 2}-C_{\varepsilon 0})}{(3C_{\varepsilon 0}-3)n+(4C_{\varepsilon 0}-3)} \times K_I, \\[4mm] \kappa = \dfrac{1}{2} \dfrac{(n+2)(C_{\varepsilon 2}-C_{\varepsilon 0})}{(3C_{\varepsilon 0}-3)n+(4C_{\varepsilon 0}-3)}. \end{cases}$$

[7] Ⓤ The limiting expression for $\mathcal{Y}_n$ here corresponds to the free fall of a simple body in vacuum. Actually, free fall in a mixing zone also involves the upward motion of the light fluid and the spreading of the kinetic energy over an ever wider layer of TMZ matter [54,55]. It is rigorously obtained by setting $K_D = K_I$ into (4.5), (4.6a) and (4.7) or using Ramshaw's model without drag [50]. In the self-similar regime the growth coefficient is now $\mathcal{Y}_n = 2/(3n+4)(n+2)$. This produces a somewhat tighter constraint on DAM's k–ε model coefficients but has no other impact in this report. It was corrected in later works [54,55].

the boundary of the constraints is greater than uncertainties of experiments and "0D" approximations (discussed in Sect. 5.2). The set of coefficients that is compatible with the constraints and best approaches the growth of $L(t)$ is:

$$\begin{cases} C_{\varepsilon 0} = \dfrac{6\sigma_c + C_{\varepsilon 2}C_\mu}{6\sigma_c + C_\mu} \approx 1.02 \\ \sigma_\varrho = \sigma_c \approx .7 \end{cases} \implies \begin{cases} \mathcal{Y}_{k\varepsilon} \approx .040 \\ \kappa \approx 1.03 \end{cases} \tag{5.17}$$

We note that the condition $\sigma_\varrho = \sigma_c$ has already been obtained by various authors based on considerations of either energy balance as here [100], or of consistency of transport in a mixture [139, 140].

5.3.3 Conclusions on DAM's k–ε Model

In conclusion, the behavior of DAM's k–ε model for an incompressible, self-similar, developed Rayleigh–Taylor instability in the limit of zero Atwood number can be summarized as follows (see Fig. 5.2):[8]

1. The coefficient σ_ϱ controls the input energy in the mixing layer, K_I, which is inversely proportional to σ_ϱ and is correctly reconstructed when $\sigma_\varrho = \sigma_c$ ($\approx .7$).
2. $C_{\varepsilon 0}$ controls the turbulence structure, characterized by the von Kármán number, $\kappa = \Lambda_i/L$. κ decreases as a function of $C_{\varepsilon 0}$, but never takes values close to those observed in direct simulations at constant acceleration, although it has permissible values around $C_{\varepsilon 0} \approx 3/2$.
3. The growth rate of the mixing zone is defined by the so-called Youngs constant, $\mathcal{Y}$, which depends on σ_ϱ and $C_{\varepsilon 0}$ according to the expression in Table 5.1, p. 52. Depending on whether $C_{\varepsilon 0} < 1$ or $C_{\varepsilon 0} > 1$, $\mathcal{Y}$ depends mainly on $C_{\varepsilon 0}$ or σ_ϱ, respectively.
4. The set of recommended values, $\sigma_\varrho = 2$ and $C_{\varepsilon 0} = .85$, underestimates the growth rate of the mixing zone by a factor of 2 and the energy balance by a factor of 3, and overestimates the von Kármán number by a factor of 14.
5. The robustness of the model can be estimated by its capacity to reconstruct certain physical conditions of a self-similar Rayleigh–Taylor instability with variable acceleration as $\Gamma \propto t^n$, although the exact experimental characteristics of this flow are not known.
6. $C_{\varepsilon 0}$ controls the existence, stability, and realizability of turbulent fluxes in RT flows with accelerations as t^n. Existence is ensured when $C_{\varepsilon 0} > 1$. When $C_{\varepsilon 0} < 1$, there is a critical exponent n_c above which self-similar solutions are not physical ($n_c = 0$ for $C_{\varepsilon 0} = 3/4$). The realizability of turbulent fluxes is ensured when $C_{\varepsilon 0} > 1.02$ approximately.

---

[8] Ⓒ New developments on SSVARTs [54,55] discussed in Sect. 9.3 have noticeably expanded points 5 to 10 of the conclusion. Stability is now ensured only if $C_{\varepsilon 0} > 3/2$.

7. In flows with accelerations as t^n, the permissible combinations of σ_ϱ and $C_{\varepsilon 0}$ are restricted to the domain defined by inequalities (5.15) and (5.16), which are associated, respectively, with compliance with the second law of thermodynamics and with a growth rate slower than free fall. When $\sigma_\varrho \geq \sigma_c = .7$ and $C_{\varepsilon 0} \geq 1.02$, both conditions are met.

8. There is no combination of coefficients σ_ϱ and $C_{\varepsilon 0}$ that reconstructs the experimental growth rate of the mixing zone at $\mathcal{Y} \approx .12$ while ensuring realizability and compliance with the second law in all flows with accelerations as t^n.

9. Although the set of recommended values $\sigma_\varrho = 2$ and $C_{\varepsilon 0} = .85$ conforms to the second law, it results in nonrealizability of solutions with accelerations as t^n when $n > n_c \approx .89$.

10. Instead of the present set of recommended values, we can propose the following coefficients:

σ_ϱ	$C_{\varepsilon 0}$
.7	1.02

which are compatible with the constraints of items 6 and 7 above, but yield a Youngs constant of $\mathcal{Y} \approx .40$, about a factor of 3 below the experimental value (while the von Kármán number is overestimated by about a factor of 8). However, model stability is not guaranteed with respect to any erratic evolution of acceleration, and thus $C_{\varepsilon 0}$ may have to be increased in some applications.

11. The constraints of energy balance and reasonable value of the von Kármán number, at $\kappa \approx .5$, result in $\sigma_\varrho = .7$ and $C_{\varepsilon 0} \approx 3/2$. Mixing zone growth is then one hundred times too slow and is associated with a directed kinetic energy balance similar to that of the Kelvin–Helmholtz instability (see Table 4.1, p. 38). This shows the inability of a k–ε type model to capture by turbulent *diffusion* the Rayleigh–Taylor mixing zone growth, which is a *directed* transport phenomenon.

6

Reconstruction by AWE's Two-Fluid Model of Developed Mixing Instabilities

6.1 "0D" Reconstruction by AWE's Two-Fluid Model of Basic Self-Similar Instabilities

6.1.1 General Remarks, Drag Characteristic Length Scale

The self-similar incompressible "0D" approach can be used to estimate the performance of AWE's two-fluid model in the three cases of simple mixing layers. In contrast to what was shown with the single-fluid k–ε model, the KH case is now singular compared to the RT and RM cases and will be discussed last. A 1D analytic study of the RT case has been published based on a slightly simplified version of the model [108, 109].

An important element shared by the three cases is that the "0D" evolution equation for the drag characteristic length, $\Lambda_d = \langle \lambda_d \rangle$ cannot be deduced from (3.32) by the standard projection procedure described in Sect. 4.4. Indeed, the directed transport term is not conservative, and contains velocity singularities at the mixing zone edges. Thus the "0D" projection must be adapted by assuming that λ_d is practically uniform across the mixing zone: this has actually been observed [33, 108. 109], and is supported by the presence of the turbulent diffusion term, as well as by the nature of the production term, which is proportional to δU_x (also uniform across the TMZ). A simple approach is thus to make Λ_d identical to the value of λ_d at the center of the mixing zone, where it is considered uniform, and then (3.32) simply yields:[1]

$$\frac{\mathrm{d}}{\mathrm{d}t}\Lambda_d = -\delta U_x \ . \tag{6.1}$$

The main characteristic of the two-fluid approach is to describe by means of evolution equations the *two* momenta associated with $U_x^{\pm}$. Thus there are several ways of realizing a "0D" projection to obtain the evolution equation of

---

[1] We note that the production terms of λ_d due to shear cancel one another for plane symmetric situations in (3.32). even in the KH case.

A. Llor: *Statistical Hydrodynamic Models for Developed Mixing Instability Flows*,
Lect. Notes Phys. **681**, 63 74 (2005)
www.springerlink.com

the directed kinetic energy, K_D, which is *the* relevant parameter presented in Sect. 4.2. Here we have decided to make a "0D" projection only of the directed kinetic energy equation as deduced from the general momentum equations. In this way we ensure the conservation of total energy, as well as the consistency with the general approach presented in Sect. 4.2.

6.1.2 Rayleigh–Taylor and Richtmyer–Meshkov Cases

Comparing (6.1) with the equation of L in (4.5), which is rigorously valid for the RT and RM cases, we obtain a general property of the "0D" version of Awe's two-fluid model: in any self-similar flow, Λ_d and L have a constant ratio:

$$\Lambda_d = \frac{L}{2} \implies \frac{\Lambda_i}{L} = \kappa_b = \frac{C_i}{2C_\mu} \quad \text{according to (3.33d)}. \tag{6.2}$$

With the coefficient values recommended in Table 3.1, p. 27, the two-fluid von Kármán number is thus $\kappa \approx .58$ for any plane self-similar flow reconstructed by Awe's two-fluid model. This ensures the realizability of the turbulent fluxes according to condition (5.8), which is still applicable with the adaptation $\sigma_c = .5$ and $\kappa \to \kappa_b$. Hereinafter, we shall replace Λ_d with $L/2$ in the "0D" analysis.

Assuming incompressibility and a vanishing Atwood number, the directed kinetic energy in a "0D" mixing zone of the RT or RM type is expressed as:

$$\bar{\varrho} k_d = (\alpha^+ \varrho^+ U_x^+ U_x^+ + \alpha^- \varrho^- U_x^- U_x^-)/2 . \tag{6.3}$$

Adding the kinetic energy equations in (3.14), we obtain from relation (4.3) and the Reynolds tensor closure presented in item 10, p. 25:

$$\frac{\partial}{\partial t}(\bar{\varrho}\, k_d) + \left(\alpha^+ \varrho^+ (U_x^+)^3/2 + \alpha^- \varrho^- (U_x^-)^3/2\right)_{,x}$$
$$= -\alpha^+ \alpha^- \frac{\varrho^+ - \varrho^-}{\bar{\varrho}} (R_{xj})_{,j}\delta U_x$$
$$- D_x^* \delta U_x + \alpha^+ \alpha^- (\varrho^+ - \varrho^-) g_x \delta U_x . \tag{6.4}$$

Integrating this equation over the mixing zone thickness and applying the "0D" approximations presented in Sect. 4.4 yields the equation in $K_D = \langle k_d \rangle$. The equation in $K_B = \langle k_b \rangle = K - K_D$ in (3.13) is obtained in the same way as presented in Sect. 5.2 for Dam's k–ε model, and the production terms by the *modeled* Reynolds stresses are here also ignored (see footnote 3, p. 51). Regarding the input energy K_I, its source term appears identical to the production term of K_D (the ambiguity justifying the introduction of K_I^* for Dam's k–ε model is not present). Finally, in the RT and RM cases, the drag D_x^* is readily expressed from $\delta U_x = -2\sqrt{3\,K_D}$, K_B, and L using (3.30). After making all the calculations, we thus obtain the system (using $\Gamma = 0$ for the Richtmyer–Meshkov instability):

$$\begin{cases} \dfrac{d}{dt}L = 4\sqrt{3\,K_D}, \\[2mm] \dfrac{d}{dt}K_I = -\dfrac{dL}{L\,dt}K_I \qquad\quad +\dfrac{2}{3}A\Gamma\sqrt{3\,K_D}, \\[2mm] \dfrac{d}{dt}K_D = -\dfrac{dL}{L\,dt}K_D - \Pi_D + \dfrac{2}{3}A\Gamma\sqrt{3\,K_D}, \\[2mm] \dfrac{d}{dt}K_B = -\dfrac{dL}{L\,dt}K_B + \Pi_D - \dfrac{2C_\mu}{C_i}\sqrt{\zeta}\,\dfrac{K_B^{3/2}}{L}, \end{cases} \qquad (6.5)$$

where the production term Π_D is deduced from item 12, Sect. 3.2, using the "0D" approximations in Sect. 4.4:

$$\Pi_D = \left\langle \frac{D_x^* \delta U_x}{\varrho} \right\rangle$$
$$= \left[\frac{C_d}{2L} S_K \left(2\sqrt{3\,K_D} - 4C_i\sqrt{\zeta K_B} \right)^2 + \frac{d}{dt}\frac{\sqrt{3\,K_D}}{4} + \frac{3\,K_D}{2L} \right] \frac{2\sqrt{3\,K_D}}{\zeta}, \quad (6.6)$$

S_K being here the sign of $2\sqrt{3\,K_D} - 4C_i\sqrt{K_B}$. Hereinafter we assume that $\zeta = 3/2$.

For self-similar RT and RM flows, we now just need to introduce the appropriate power laws in t for K_D, K_B, and L into the differential system (6.5). The substitution is trivial for the linear equations in L and K_I, and makes it possible to express K_I and K_D as functions of the growth of $L(t)$; naturally, the relations obtained are *identical* to those obtained in Sect. 4.2, because these two quantities were defined using the two-fluid approach. In contrast to the case of DAM's k-ε model, the balance of energy dissipated as internal energy is thus always correct, and always complies with the second law of thermodynamics. For the last two equations, in K_D and K_B, we replace K_B by:

$$\mathcal{U} = \sqrt{\frac{K_B}{K_D}}\,. \qquad (6.7)$$

To make Π_D explicit, one must first choose a value of the sign S_K (a priori $+1$), which must be consistent with the values of K_D and K_B that will be eventually obtained. It must then be checked that:

$$\frac{K_B}{K_D} < \frac{3}{4C_i^2} \approx 45.4 \ \text{ if } \ S_K = +1\,. \qquad (6.8)$$

From there, lengthy and tedious though straightforward calculations reduce the equations in K_D and K_B to:

- an algebraic equation in $\mathcal{U}$ as a function of coefficients C_μ, C_d, and C_i, of third or fourth degree for the RT and RM cases, respectively, and

- an explicit formula giving the growth rate as a function of $\mathcal{U}$ and the coefficients.

In both cases, only one of the roots of the equation in $\mathcal{U}$ is real, positive, and compatible with condition (6.8). In the RM case, considering the large ratio K_B/K_D, we can obtain a good approximation, simple and accurate to within 1%, by considering the equation in $K = K_B + K_D$: it is identical to that of K_B in (6.5), where the term Π_D would be disregarded because $\Gamma = 0.^2$ $\mathcal{U}$ is then a solution of a second-degree equation. The results are summarized in Table 6.1.

6.1.3 Kelvin–Helmholtz Case

In contrast to the k–ε model discussed in Sect. 5.2, two-fluid models such as AWE's introduce a radically different behavior for the KH instability compared with the RT and RM instabilities:

1. The directed kinetic energy has two components: normal and tangential to the mixing layer. To be entirely rigorous, these should be deduced from separate momentum equations, but the drag expression then becomes rather cumbersome. However, the velocity difference between the fluids is essentially normal, because it includes turbulent dispersion (3.31), which is related to the concentration gradient.
2. For practical purposes, the normal directed kinetic energy, K_D is small compared with the turbulent kinetic energy K_B (of the order of 1%, see Table 4.1, p. 38) and corresponds to a relative velocity of the fluids that is close to the velocity of turbulent dispersion (see condition (6.8)).
3. The growth of the *mixing* zone is limited by the growth of the *shear* zone due to turbulent diffusion of the transverse momentum. The equation in $L(t)$ in (6.5) thus contains $\sqrt{K_B}$ rather than $\sqrt{K_D}$ (as for the 0D reduction of DAM's k–ε model in (5.6)). The concentration profile is no longer linear, and its slope at the center is a factor of $\zeta' \approx 1.25$ less than that of the transverse velocity profile (see C). As a result, expression (6.2) must be modified.

The sum of the transverse momentum equations leads to the same equation as for a single-fluid model (2.6), *provided the difference between U_y^+ and U_y^- is disregarded*, as is permitted by item 2 above. Assuming incompressibility and zero Atwood number, AWE's model thus yields the diffusion coefficient for U_y as:

$$C_i \lambda_d \sqrt{\bar{k}} \,. \tag{6.9}$$

Based on the reasoning that led to (5.2), we obtain from this the evolution equation for L in system (6.10) below, which must also include the equation

---

[2] This approximation is not acceptable in the RT case, because the K_B/K_D ratio is close to 1.

of Λ_d in (6.1) in accordance with item 3 above. Thus we note that L and $2\Lambda_d$ represent the lengths of the respective gradients of transverse momentum and concentration. For the turbulent kinetic energy production terms, we can disregard the drag (dissipation of K_D) relative to the Reynolds stresses, also per item 2 above, and based on the reasoning that led to (5.3), we thus obtain the expressions of the equations for K_I and K_B in (6.10) (Π_D is disregarded in the equation of K_B as in the RM case above). As for the equation in K_D, we must note the absence of the source term, which is present in the RT case in (6.5): the work of the drag Π_D now represents a term of K_D *production*. In the "0D" projection of AWE's model, the KH instability is therefore described by the system:

$$
\left\{
\begin{aligned}
\frac{d}{dt}L &= 8C_i \frac{\Lambda_d}{L}\sqrt{\zeta K_B}. \\[2ex]
\frac{d}{dt}\Lambda_d &= 2\sqrt{3\,K_D}, \\[2ex]
\frac{d}{dt}K_I &= -\frac{dL}{L\,dt}K_I + C_i\Lambda_d\sqrt{\frac{K_B}{\zeta}}\left(\frac{\Delta U_y}{L}\right)^2, \\[2ex]
\frac{d}{dt}K_D &= -\frac{dL}{L\,dt}K_D \qquad\qquad\qquad - \Pi_D, \\[2ex]
\frac{d}{dt}K_B &= -\frac{dL}{L\,dt}K_B + C_i\Lambda_d\sqrt{\frac{K_B}{\zeta}}\left(\frac{\Delta U_y}{L}\right)^2 - \frac{C_\mu}{C_i}\sqrt{\zeta}\,\frac{K_B^{3/2}}{\Lambda_d}.
\end{aligned}
\right.
$$

$$\tag{6.10}$$

Taking into account note 1 above, Π_D is now written as:

$$
\Pi_D = \left[\frac{C_d}{4\Lambda_d}\mathcal{S}_K\left(2\sqrt{3\,K_D} - 4C_i\sqrt{\zeta K_B}\right)^2 \right.
$$
$$
\left. + \frac{d}{dt}\frac{\sqrt{3\,K_D}}{4} + \frac{3\,K_D}{4\,\Lambda_d}\right] 2\sqrt{3\,K_D}, \quad (6.11)
$$

as a function of the two characteristic lengths L and Λ_d.[3] Here also, by introducing appropriate power laws in t for the KH case, lengthy and tedious though straightforward calculations yield the results summarized in Table 6.1.

The results of the K_B/K_D ratio are consistent with condition (6.8) only for the choice $\mathcal{S}_K = -1$ used here (the opposite of that used in the RT and RM cases). The value obtained for the von Kármán number is compatible

---

[3] In the spirit of item 3 above and in order to simplify the expressions, K_D in equations (6.10) and (6.11) only represents the directed energy at the *center* of the mixing zone, $k_d(0)$ and not the *mean*, $\langle k_d \rangle$. Thus a reduction by a factor of the order of ζ should be applied to it.

Table 6.1. Basic geometric and energy characteristics of KH, RT, and RM turbulent mixing zones as reconstructed by AWE's "0D" simplified two-fluid model. The boxed numerical values were obtained with the standard coefficients of AWE's two-fluid model in Table 3.1. These modeled results are to be compared with the "0D" experimental data given in Table 4.1, p. 38. The approximate treatment in the KH case introduces a correction to K_D by a factor of $2^{1/2}/(\varsigma C_i u_{KH}) \approx 1.14$, which is close to the reference value of unity

Instability	KH	RT	RM
Constant	$\mathcal{X}_A = \dfrac{2\sqrt{2}C_i}{\sqrt{C_\mu}\,u_{KH}^2 + 2\sqrt{2}C_i\,u_{KH}}$ $\boxed{\approx .105}$	$\mathcal{Y}_A = \dfrac{1}{C_d\left(1-\sqrt{2}C_i\,u_{RT}\right)^2 + 3/2}$ $\boxed{\approx .106}$	$n_A = \dfrac{8C_i}{\sqrt{2}C_\mu u_{RM} + 12C_i}$ $\boxed{\approx .44}$
L	$\mathcal{X}_A \times \Delta U_y t$	$\mathcal{Y}_A \times A\Gamma t^2$	$L_0\left(\dfrac{t}{t_0}\right)^{n_A}$
K_I	$\dfrac{1}{12} \times \Delta U_y^2$	$\dfrac{\mathcal{Y}_A}{12} \times (A\Gamma t)^2$	$K_{I0}\left(\dfrac{t}{t_0}\right)^{-n_A}$
K_D	$\dfrac{4}{\varsigma\left(2+\sqrt{7/2/C_d}\right)} \times \dfrac{\mathcal{X}_A^2}{48} \times \Delta U_y^2$	$\dfrac{\mathcal{Y}_A^2}{12} \times (A\Gamma t)^2$	$\dfrac{n_A^2}{48} \times \left(\dfrac{L_0}{t_0}\right)^2 \left(\dfrac{t}{t_0}\right)^{-2(1-n_A)}$

Table 6.1. continued

Instability	KH	RT	RM
K_B/K_I	$\dfrac{2\sqrt{2}C_i}{C_\mu \mathcal{U}_{\mathrm{KH}} + 2\sqrt{2}C_i}$ $\approx .3$	$\mathcal{Y}_A\,\mathcal{U}_{\mathrm{RT}}^2(C_\mu, C_d, C_i)$ $\approx .65$	$\mathcal{U}_{\mathrm{RM}}^2 = \dfrac{\left(\sqrt{64C_d C_i^2 + 3C_\mu} - \sqrt{3C_\mu}\right)^4}{\left(64\sqrt{2}C_d C_i^3\right)^2}$ ≈ 26
K_B/K_D	$\zeta\,\mathcal{U}_{\mathrm{KH}}^2 = \zeta\left(\dfrac{2+\sqrt{7/2}/C_d}{2\sqrt{2}C_i}\right)^2$ ≈ 93	$\mathcal{U}_{\mathrm{RT}}^2(C_\mu, C_d, C_i)$ ≈ 6.2	
κ_b	$\dfrac{C_i}{C_\mu}\dfrac{1}{\sqrt{2+\sqrt{7/2}/C_d}}$ $\approx .76$		$\dfrac{C_i}{2C_\mu}$ $\approx .58$

$$\sqrt{2}C_\mu\,\mathcal{U}_{\mathrm{RT}}^3 - 8C_i(2C_d C_i^2 - 1)\mathcal{U}_{\mathrm{RT}}^2 + 16\sqrt{2}C_d C_i^2\,\mathcal{U}_{\mathrm{RT}} - 4C_i(2C_d+1) = 0\,.$$

with the realizability constraint of turbulent fluxes (5.8), and the Reynolds tensor is positive as with the k–ε model in Sect. 5.2.

6.2 Calibration of Coefficients C_i and C_d of AWE's Two-Fluid Model

6.2.1 Effect of the Coefficients on KH, RT, and RM Instabilities

A comparison of Tables 4.1, p. 38 and 6.1, p. 68 shows that the values of the various energy parameters of the KH, RT, and RM instabilities are generally well reconstructed by AWE's two-fluid model. The most pronounced differences deserve some comments:

- $\mathcal{Y}_0$ is underestimated by about 15%, but it is not possible at this level to determine the cause: either a bias of the "0D" projection or an unsuitable value of the coefficient C_d adjusted by comparison with direct simulations [33] that underestimate $\mathcal{Y}_0$;
- n_0 is overestimated by 33%, probably because the model assumes Schmidt–Prandtl numbers of $1/2$. A more realistic value of $\sigma_c = .7$ would reduce the turbulent diffusion, which is dominant in the RM case;
- the von Kármán numbers are consistently overestimated, not to much in the KH and RM cases (25% and 50%), but by as much as a factor of the order of 6 in the RT case. An adjustment of C_i would reduce these discrepancies, but would be detrimental to the reconstruction of growth rates;
- K_B/K_D is clearly overestimated by 60% in the RT case, due to a combination of $\mathcal{Y}_0$ underestimation of and K_B/K_I overestimation by about 30% (this relates to the large overestimation of κ_b).

The value of C_μ, which controls the turbulent viscosity, must retain its "universal" value as in the k–ε model. Thus the only adjustable parameters of AWE's two-fluid model are C_i and C_d, as a function of which we can draw contour lines of the growth coefficients using the formulas of Table 6.1, p. 68. The result in Fig. 6.1 shows:

- the weak dependence of coefficients $\mathcal{X}_A$ and n_A on C_d,
- the decay of $\mathcal{X}_A$ and n_A as a function of C_i,
- the decay of $\mathcal{Y}_A$ as a function of C_d,
- the growth of $\mathcal{Y}_A$ as a function of C_i.

These behaviors are quite natural: in the KH and RM cases, AWE's two-fluid model degenerates into a form of k–ℓ that no longer includes C_d (K_D can be disregarded, see Sect. 6.1), and the dissipation increases with C_i, thereby slowing the growth; in the RT case, the drag increases with C_d, resulting in a decrease of $\mathcal{Y}$.

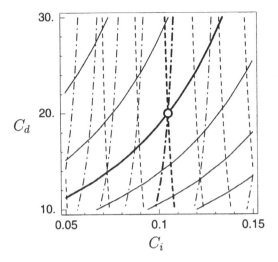

Fig. 6.1. Graphical representation in the (C_i, C_d) plane of contour lines of the growth coefficients of self-similar KH, RT, and RM instabilities obtained from the "0D" projection of AWE's two-fluid model: (*dot-and-dash lines*) constant $\mathcal{X}$, spaced by $.25 \times \mathcal{X}_A$ (KH); (*solid lines*) constant $\mathcal{Y}$, spaced by $.2 \times \mathcal{Y}_A$ (RT); (*dashed lines*) constant n, spaced by $.1 \times n_A$ (RM); (*heavy lines*) contour lines drawn from the combination of recommended model coefficients, as marked by (*empty circle*) (see Table 3.1)

The evolutions of the basic parameters for the three instabilities are shown in Fig. 6.2 as a function of C_i when C_d is set at its recommended value, and vice versa. A value of $C_d \approx 16$ makes it possible to reconstruct the Youngs constant, $\mathcal{Y}_A = \mathcal{Y}_0$. We observe that all the parameters remain physically permissible over a broad range of coefficients around their recommended values.

6.2.2 Stability and the RT Case with Self-Similar Variable Acceleration[4]

Regarding the stability of the model, we can examine, as for DAM's k–ε model, the linear stability of small perturbations and the Rayleigh–Taylor instability with self-similar acceleration varying as $\Gamma \propto t^n$, previously introduced in Sect. 5.3. The calculation of small perturbations, which is not presented here, shows stability in a wide zone of values of C_d and C_i. The behavior is more instructive for the Rayleigh–Taylor instability with variable self-similar acceleration. System (6.5) is solved by lengthy and tedious though straightforward calculations to yield:

---

[4] ⓒ New developments on SSVARTs [54,55] discussed in Sect. 9.3 have noticeably expanded the findings of this part.

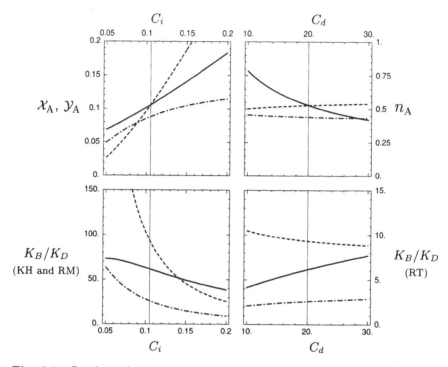

Fig. 6.2. Graphs as functions of the coefficients C_d at $C_i = .105$ (*left*), and C_i at $C_d = 20$ (*right*) of the self-similar growth coefficients (*top*) $\mathcal{X}_A$, $\mathcal{Y}_A$ (*left scale*), and n_A (*right scale*), and of the ratios of the two-fluid turbulent and directed kinetic energies, K_B/K_D (*bottom*), as obtained from the "0D" projection of AWE's two-fluid model for the three basic instabilities: (*dashed lines*) Kelvin–Helmholtz; (*solid lines*) Rayleigh–Taylor (K_B/K_D scale on the right magnified by a factor of 10); (*dot-and-dash lines*) Richtmyer–Meshkov (scale of n_A on the right reduced by a factor of 5). (*vertical lines*) mark the recommended values of the coefficients (see Table 3.1)

$$
\begin{cases}
L &= \mathcal{Y}_n \times A\Gamma t^2 \; = \; \dfrac{4/(2+n)^2}{C_d\left(1-\sqrt{2}C_i\,\mathcal{U}_n\right)^2 + 3(1+3n/4)/(2+n)} \; A\Gamma t^2, \\[2ex]
K_I &= \dfrac{(1+n/2)\mathcal{Y}_n}{12\,(1+3n/4)} \times (A\Gamma t)^2, \\[2ex]
K_D &= \dfrac{(1+n/2)^2\mathcal{Y}_n^2}{12} \times (A\Gamma t)^2, \\[2ex]
K_B &= \mathcal{U}_n^2(c_\mu, c_d, c_i) \times K_D, \\[2ex]
\kappa_b &= \dfrac{C_i}{2C_\mu},
\end{cases}
$$

$$(6.12)$$

with the equation in $\mathcal{U}_n$:

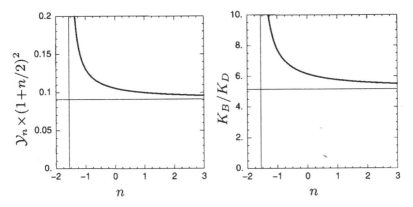

Fig. 6.3. Evolution, as a function of the exponent n, of the normalized growth coefficient of the mixing zone, $\mathcal{Y}_n \times (1+n/2)^2$, and of the ratio of the two-fluid turbulent and directed energies, K_B/K_D, for a self-similar Rayleigh–Taylor instability with acceleration varying as t^n, as obtained using AWE's two-fluid model

$$\sqrt{2}C_\mu \, \mathcal{U}_n^3 \; - \; 8C_i \left[2C_d C_i^2 - 2\frac{1+3n/4}{2+n} \right] \mathcal{U}_n^2$$
$$+ \; 16\sqrt{2}C_d C_i^2 \, \mathcal{U}_n \; - \; 4C_i \left[C_d + 2\frac{1+3n/4}{2+n} \right] \; = \; 0 \, ,$$
$$(6.13)$$

while relation (6.2), which gives the von Kármán number, remains of course valid in the case of variable acceleration.

As shown in Fig. 6.3, the characteristic parameters $\mathcal{Y}_n \times (1+n/2)^2$ and K_B/K_D do remain finite for all $n > -1$. The observed asymptotic behaviors are obtained by taking the limit $n \to \infty$ of the equations (6.12) and (6.13) which, with the recommended coefficient values, yields the numerical results:

$$\mathcal{Y}_\infty \approx .091 \times (1+n/2)^2 \quad \text{and} \quad (K_B/K_D)_\infty \approx 5.17 \, . \qquad (6.14)$$

Compared with the case of constant acceleration at $n = 0$, we observe between $n = -1$ and $n = \infty$ variations of about $+20\%$ to -20% in the normalized

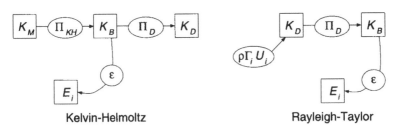

Fig. 6.4. Diagram of energy transfers in Kelvin–Helmholtz and Rayleigh–Taylor instabilities according to AWE's two-fluid model. Compare this diagram with that of Fig. 5.1, p. 50

growth rate and the K_B/K_D ratio. Below $n = -1$, these values diverge for $n = n_0 \approx -1.56$. The normalization of $\mathcal{Y}_n$, as $(1+n/2)^{-2}$ is related to its expression in (6.12) and will be justified in Sect. 7.4 by formula (7.16), p. 91.

This "quiet" behavior of AWE's two-fluid model, compared with DAM's k–ε model, is due to a consistent structure of the exchange paths between the potential, two-fluid turbulent, directed, and internal energies as diagrammed in Fig. 6.4. However, this is no guarantee that the model is *accurate* for variable acceleration flows, for which few numerical experimental or theoretical values are available.

Summary of Part II

In this part, models have been applied to basic flows and their predictions compared to the known experimental results.

All the turbulent mixing flows considered here, KH (Kelvin–Helmholtz), RT (Rayleigh–Taylor), RM (Richtmyer–Meshkov), and more marginally SS-VARTs (self-similar variable acceleration Rayleigh–Taylor) were self-similar and reduced to their *"0D" averages* in the limit of zero Atwood number. The term "0D" average designates average over the TMZ (turbulent mixing zone), which, by symmetry, can be readily calculated at zero Atwood number. This approach not only simplifies dramatically all model analyses (which become algebraically tractable), but also provides useful insights on the underlying physical processes, at the expense of approximations with marginal impact. A systematic procedure for "0D" reduction was given.

Four basic "0D" averaged quantities have been considered here (Table 4.1, p. 38): the growth rate of the TMZ, the directed energy, the turbulent energy and the *von Kármán number* (ratio of integral length scale to the TMZ width). These quantities are not physically independent (only two are independent) but may decouple in some models, such as DAM's k-ε.

Most noticeable is the von Kármán number, which appears to be *significantly smaller in RT* than in more common KH and RM flows, by almost an order of magnitude. This unique feature motivated the coining of a specific name by analogy with the Kármán *constant*.

Analytical '0D" responses of the two models retained in this study were obtained (Tables 5.1, p. 52 and 6.1, p. 68). Although growth rates and various other parameters are reasonably captured for all instabilities, the von Kármán number of the RT flow is significantly overestimated in both models.

DAM's k-ε also appears to hinge on an unphysical energy path which, in some cases, is incompatible with the second law of thermodynamics. As a consequence the response of the model under variable acceleration can become unphysical (regardless of the availability of experimental data): for instance, growth of the TMZ as restituted by DAM's k-ε is found to *overrun the free fall limit* for some SSVARTs. AWE's two-fluid model is always acceptable according to this criterion, although its accuracy could be debated.

SSVARTs represent a form of decomposition basis for general variable acceleration flows, and as such they provide useful test flows for models (as examplified in the case of DAM's k–ε).

Comparative Assessment of Models,
New Development Approaches

7

Comparison of Single- and Two-Fluid Approaches

7.1 Summary of "0D" Reconstructions of the Three Instabilities by the Single- and Two-Fluid Models

The results of the "0D" analyses gathered in Tables 5.1 and 6.1 and discussed in Sects. 5.3 and 6.2 provide the basis for comparing DAM's k–ε and AWE's two-fluid models as outlined in Table 7.1. No further comments will be given on this comparison except for one specific point that is crucial to modeling: the energy dissipation circuits. The representation of these circuits in Table 7.1 is rearranged and simplified from Figs. 5.1 and 6.4.

The directed and turbulent kinetic energies, K_D and K, though both of kinetic nature, have different status regarding entropy. Under the reasonable assumption of heterogeneous mixing (see Sect. 9.2), after a Rayleigh–Taylor instability there could readily be a transient retransfer of a sizable portion of K_D back into potential energy by inverting the acceleration (demixing). By contrast, the same conversion of K could not be imagined for a Kelvin–Helmholtz instability, even though in certain specific flows it is possible to "redirect" a *small* fraction of K: for example, in localized separation and recirculation zones of boundary layers, and also more generally in the KH case, where K_D is fed by K, as clearly shown in the reconstruction by AWE's two-fluid model (see Sect. 6.1).

The circuit of AWE's two-fluid model correctly captures the following properties:

- two-directional exchanges between E_{RT} and K_D, capturing both mixing and demixing phases;
- two-directional exchanges between K_D and K_B, capturing KH type situations ($K_B \rightarrow K_D$) and RT and RM situations ($K_D \rightarrow K_B$);
- purely dissipative exchange between K_B and E_I, compatible with the second law of thermodynamics.

By contrast, in DAM's k–ε model:

A. Llor: *Statistical Hydrodynamic Models for Developed Mixing Instability Flows*,
Lect. Notes Phys. **681**, 79–92 (2005)
www.springerlink.com   © Springer-Verlag Berlin Heidelberg 2005

Table 7.1. Comparative table summarizing the main differences between DAM's k–ε model and AWE's two-fluid model applied to KH, RT, and RM instabilities in the incompressible limit at zero Atwood number

Property	DAM's k–ε model	AWE's two-fluid model
Reconstruction of **KH and RM** instabilities	**Correct:** The model was designed to reconstruct two flows very closely related to KH and RM: boundary layer turbulence and grid turbulence decay.	**Correct:** For these instabilities, the model degenerates into a form of k–ℓ.
Reconstruction of **RT instability** with constant or variable acceleration	**Incorrect:** No choice of parameters can *both* yield the experimental growth rates and keep the model stable when acceleration varies. The reconstructed growth rate is at best of the order of 30% of the observed rate.	**Correct:** But the von Kármán number is overestimated by a factor of 6 because there is just one characteristic length scale.
Relative flux between fluids (velocity difference), or directed kinetic energy, K_D	**Implicit:** Entirely contained in turbulent diffusion. K_D is included in the turbulent kinetic energy, K, from which it is not distinguished.	**Explicit:** Momentum equations of the fluids with drag term. Closure of drag makes it possible to capture behaviors dominated by either diffusive effects (KH, RM) or directed effects (RT).
Dissipation circuit of energy sources of instabilities, E_{KH} and E_{RT}, into internal energy, E_I	$$\begin{array}{cc} E_{\mathrm{KH}} & E_{\mathrm{RT}} \\ \downarrow & \uparrow \\ K & \leftrightarrow\ E_I \\ (=K_B+K_D) & \end{array}$$	$$\begin{array}{cc} E_{\mathrm{RT}} & E_{\mathrm{KH}} \\ \uparrow & \downarrow \\ K_D\ \leftrightarrow & K_B \quad \to\ E_I \\ & (=K-K_D) \end{array}$$
Distinctive features of the dissipative circuit in RT	**Possible inconsistencies depending on coefficients:** • growth faster than free fall, • violation of second law.	**Consistency over wide range of coefficients:** Ensured by the inclusion of evolution equations for all energies, the balance of which is always dissipative.

- E_{RT} is coupled with E_I in a way that is necessarily two-directional (via the mean kinetic energy, see Fig. 5.1) through the "reversible" term for the work of pressure;
- E_I is coupled with $K = K_B + K_D$ also in a two-directional way to allow both dissipation and transfer of E_{RT} into $K_B + K_D$.

Given that in the incompressible limit, E_I reduces to a receptacle of dissipation, DAM's $k\text{-}\varepsilon$ model thus leaves open possibilities for thermodynamic inconsistencies if precautions are not taken.

The following two sections analyze how these dissipation circuits are embodied in the statistical equations, starting with a general single-fluid approach, then establishing the link between the single- and two-fluid approaches. This leads to suggestions for improving the models: extended single-fluid models and degenerate two-fluid models.

7.2 About the "Rayleigh–Taylor" Source Term in Single-Fluid Models[1]

The Boussinesq–Reynolds closure applied to the turbulent mass flux $\overline{u_i''}$ in the so-called "Rayleigh–Taylor production" term in (2.17), π_{RT}, might seem natural and consistent with the general methods of turbulence modeling. The "intuitive" argument usually provided a posteriori to justify it [142, §2.1] is an analogy with the baroclinic vorticity production term, $\nabla p \wedge \nabla \varrho$, "averaged" over a random distribution of the fluids in the mixing zone. Although reassuring by its simplicity, intuitive character, quick implementation into codes, and "suitable" performance for reconstructing certain experiments, this closure introduces numerous difficulties, of which we can cite the following:

1. The model (2.17) was first used outside of the DAM to treat compressible single-fluid flows other than mixing instabilities (boundary layers, etc.) under a set of fairly restrictive assumptions [61, 63, 79, 80, §16.3] (ideal gas, homogeneity of stagnation thermodynamic conditions, etc.). More recently, other authors have proposed, without changing the general form of the closure of $\overline{u_i''}$, that the coefficient σ_ϱ associated with it depends on the turbulent Mach number $\mathrm{M}_t = \tilde{k}^{1/2}/c_s$ [63, 143, 144, §3.3]: thus $\overline{u_i''}$ is canceled in the incompressible limit.

---

[1] Ⓤ This section was the very first written in this report, in 1998, and the importance of the related entropy aspects was not fully realized at this stage (although some specific entropy considerations appear in other places as E). However, interesting constraints should be deduced from general analysis of entropy equations in turbulence models [141]. An other related aspect is the pressure decomposition in the two-fluid approach of Sect. 3.1 which should have been given in a more conventional and consistent way, with different fluid pressures $P^{\pm}$.

2. When the product $(\bar{p})_{,i}(\bar{\varrho})_{,i}$ is positive (demixing, stratified turbulence, etc.), the destruction of turbulence often seems too strong to permit suitable reconstructions. Thus modelers have introduced limiters, and even the complete cancellation of the "Rayleigh–Taylor term" based on purely empirical basis instead of physical considerations or realizability constraints [3–6, 8–11, 13–17].

3. As was discussed in Sect. 5.3, Rayleigh–Taylor instabilities are not reconstructed in a way that is both precise and stable, and certain choices of coefficients lead to violations of the second law of thermodynamics.

The theoretical principle underlying all these difficulties is that the modeling of $\overline{u_i''}$, which appears in the equation of $\overline{\varrho k}$ in (2.11), cannot be dissociated from that of the corresponding terms in the other energy equations. In the Boussinesq–Reynolds form, the "Rayleigh–Taylor term" creates turbulent kinetic energy from the reservoir of *internal energy*, hence the problems of compatibility with the second law of thermodynamics. This was shown in the "0D" reduction of the model in Sect. 5.2 and Fig. 5.1, and was illustrated for the self-similar RT instability in Sect. 5.3. Therefore, we must take a general look at ways to remedy these deficiencies.

Before attempting any modeling, we must note that the choice of a decomposition for the various terms in the statistical equations is not always neutral, even though mathematically exact, because it corresponds to *implicit physical hypotheses* that orient the closures and noticeably modify the modeled flow. The first two trivial decompositions of the "Sound and Rayleigh–Taylor" terms in the energy balance (2.11) are easily demonstrated [142, §2.2]:

$$
\begin{array}{l}
\frac{\partial}{\partial t}\overline{\varrho e}\ \cdots\cdots \\[4pt]
\frac{\partial}{\partial t}\overline{\varrho k}\ \cdots\cdots \\[4pt]
\frac{\partial}{\partial t}\overline{\varrho U^2/2}\ \cdots \\[4pt]
\frac{\partial}{\partial t}\overline{\varrho f}\ \cdots\cdots
\end{array}
\begin{bmatrix}
\text{Sound and RT} \\
-\ \bar{p}\,(\overline{u_i})_{,i} \\
-\ (\bar{p})_{,i}\,\overline{u_i''} \\
-\ (\bar{p})_{,i}\,U_i \\
-\ (\overline{p\,u_i})_{,i}
\end{bmatrix}
=
\begin{bmatrix}
\text{1: Reynolds average} \\
-\ \bar{p}\,(\overline{u_i})_{,i} \qquad \cdot \\
\cdot \qquad -\ (\bar{p})_{,i}\,\overline{u_i''} \\
-\ (\bar{p})_{,i}\,\overline{u_i}+(\bar{p})_{,i}\,\overline{u_i''} \\
-\ (\overline{p\,u_i})_{,i} \qquad \cdot
\end{bmatrix}
=
\begin{bmatrix}
\text{2: Favre average} \\
-\ \bar{p}(U_i)_{,i}-\bar{p}\,(\overline{u_i''})_{,i} \\
\cdot \qquad -\ (\bar{p})_{,i}\,\overline{u_i''} \\
-\ (\bar{p})_{,i}\,U_i \qquad \cdot \\
-\ (\bar{p}\,U_i)_{,i}-(\overline{p\,u_i''})_{,i}
\end{bmatrix}.
$$

$$(7.1)$$

The former emphasizes the Reynolds averaged overall motion, and the $\overline{u_i''}$ term is thus associated with the exchange between the mean and turbulent kinetic energies, because exchange between internal and turbulent kinetic energies is now forbidden. The latter corresponds to the Favre averaged overall motion, and thus produces the exact inverse to the former, forbidding exchange between mean and turbulent kinetic energies.

Naturally, these two extremes are meaningless, because exchanges with internal energy result from compressions due to both the mean motion and the turbulent motion. Although the work of the Favre averaged pressure $\bar{p}(U_i)_{,i}$ is not explicitly present in (2.11), its *dual term* $(\bar{p})_{,i}\,U_i$ is indeed there, and to obtain the distribution of transfers, we must analyze its nature. It can be written:

$$(U_i)_{,i} = \left(\overline{\frac{\varrho\, u_i}{\overline\varrho}} \right)_{,i} = \overline{\frac{\varrho\, u_{i,i}}{\overline\varrho}} + \overline{\frac{(\varrho)_{,i}\, u_i}{\overline\varrho}} - \overline{\frac{(\varrho)_{,i}\, \varrho\, u_i}{\overline\varrho^2}} = \boxed{\overline{\frac{\varrho\, u_{i,i}}{\overline\varrho}}} + \boxed{\overline{\frac{(\varrho)_{,i}\, u_i''}{\overline\varrho}}} , \quad (7.2)$$

where

$$\overline{\frac{\varrho\, u_{i,i}}{\overline\varrho}} = -\frac{1}{\overline\varrho} \overline{\left(\frac{\mathrm{d}\varrho}{\mathrm{d}t} \right)} , \qquad (7.3)$$

which shows the coexistence of two contributions of very different types in $\mathrm{div}\, U$, the first due to fluid compression (Favre average of $\mathrm{div}\, u$) and the second due to non-uniform density (correlation between density gradients and velocity fluctuations). Using (7.2), we can thus provide a third decomposition of the "Sound and Rayleigh–Taylor" terms:

$$
\begin{array}{l}
\frac{\partial}{\partial t}\overline{\varrho e} \ldots\ldots \\[4pt]
\frac{\partial}{\partial t}\overline{\varrho k} \ldots\ldots \\[4pt]
\frac{\partial}{\partial t}\overline{\varrho U^2/2} \ldots \\[4pt]
\frac{\partial}{\partial t}\overline{\varrho f} \ldots\ldots
\end{array}
\left[
\begin{array}{l}
\text{Sound and RT} \\
-\ \overline{p}\,(\overline{u_i})_{,i} \\
-\ (\overline{p})_{,i}\, \overline{u_i''} \\
-\ (\overline{p})_{,i}\, U_i \\
-\ (\overline{p\, u_i})_{,i}
\end{array}
\right]
\tag{7.4}
$$

$$
=
\left[
\begin{array}{ccc}
\multicolumn{3}{c}{\text{3: Compressions and density heterogeneities}} \\
-\overline{p}\,\overline{\varrho\,(u_i)_{,i}}/\overline\varrho & \cdot & +\overline{p}\,\overline{\varrho'\,(u_i'')_{,i}}/\overline\varrho \quad \cdot \\
 & -\overline{p}\,\overline{(\varrho)_{,i}\, u_i''}/\overline\varrho - \overline{p}\,\overline{\varrho'\,(u_i'')_{,i}}/\overline\varrho - (\overline{p\,u_i''})_{,i} & \cdot \\
-\ (\overline{p})_{,i}\, U_i & \cdot & \cdot \\
-\ (\overline{p}\, U_i)_{,i} & \cdot & -\,(\overline{p\,u_i''})_{,i}
\end{array}
\right] \cdot
$$

The terms in $\overline{\varrho\, u_{i,i}}$ and $\overline{(\varrho)_{,i}\, u_i''}$ in (7.2) are necessarily associated with transfers to internal and turbulent energy, respectively. This makes it possible to properly account for the two limit behaviors: incompressible fluids or homogeneous density. Then the overall balance shows that *the turbulent mass flux disappears completely due to exchanges between different types of energies* and is reduced to a turbulent kinetic energy flux term. Naturally, we now see several new terms that require as many specific modelings and associated assumptions. Incidentally, the term $\overline{p}\,\overline{u_i''}$ is grouped with the term $\overline{p'u_i''}$ in (2.11) to give the flux by correlations between pressure and velocity fluctuations, $\overline{p\,u_i''}$, which is classically separated and modeled in the equation of turbulent kinetic energy of an incompressible homogeneous fluid [61, §6.3].

We note that the correlation $\overline{(\varrho)_{,i}\, u_i''}$ between density gradients and velocity fluctuations is central to the exchanges between mean and turbulent kinetic energies in (7.4). The modeling of this correlation can by no means be reduced to a simple closure on kinematic criteria, unlike what has been done abusively for the turbulent mass flux, except when invoking specific assumptions of *uniformity of the equation of state and of the stagnation density* which obviously do not apply to Rayleigh–Taylor instabilities [61,63,79,80, §16.3]. In the case of mixing instabilities, modeling is greatly simplified by use of the two-fluid description, as is shown in Sect. 7.3 below.

The "Sound and RT" terminology used so far to designate the average pressure terms in the statistical equations can now be justified from (7.4): "Sound" stands for all the energy exchanges involving compression (i.e. with internal energy) whether slowly varying, propagative (waves, shocks) or random (noise), whereas "RT" stands for the incompressible energy exchanges (i.e. between mean and turbulent kinetic energies) thus involving density gradients only. The latter is the genuine production term of turbulent kinetic energy in the Rayleigh–Taylor instability.

7.3 Two-fluid Realizability of Single-Fluid Closures of Turbulent Fluxes

The constraints introduced by the single-fluid description in the case of a mixing instability between different fluids can be clearly analyzed by combining the two-fluid statistical equations into their "common mode." Decomposing the single-fluid quantities into two-fluid quantities and taking into account relations (3.6) and (3.7a), we have for any per mass quantity a:

$$
\begin{aligned}
\overline{\varrho a u_i''} &= \overline{\varrho a u_i} - \overline{\varrho a} U_i \\
&= \overline{c^+ \varrho a u_i^+} + \overline{c^+ \varrho a} U_i^+ + \overline{c^- \varrho a u_i^-} + \overline{c^- \varrho a} U_i^- - \overline{\varrho a} \frac{\overline{c^+ \varrho}}{\overline{\varrho}} U_i^+ - \overline{\varrho a} \frac{\overline{c^- \varrho}}{\overline{\varrho}} U_i^- \\
&= \overline{c^+ \varrho a u_i^+} + \overline{c^- \varrho a u_i^-} + \frac{\overline{c^+ \varrho a}\, \overline{c^- \varrho} - \overline{c^- \varrho a}\, \overline{c^+ \varrho}}{\overline{\varrho}} \delta U_i \\
&= \overline{c^+ \varrho a u_i^+} + \overline{c^- \varrho a u_i^-} + C^+ C^- \overline{\varrho} (A^+ - A^-) \delta U_i ,
\end{aligned}
\tag{7.5}
$$

where the $A^\pm$ are the Favre averaged per mass quantities for each fluid as defined in (3.4).

The total turbulent flux in (7.5) thus comprises the turbulent fluxes for each fluid, as well as a *directed* contribution related to the *interpenetration velocity of the fluids* $\delta U = U^+ - U^-$. For the quantities considered here, namely, concentration, fluctuating velocity, mass, and internal energy, we thus obtain:

$$
\begin{aligned}
\overline{\varrho c^\pm u_i''} &= \pm C^+ C^- \overline{\varrho}\, \delta U_i , \\[2mm]
\overline{r_{ij}} &= C^+ C^- \overline{\varrho}\, \delta U_i\, \delta U_j + \overline{c^+ r_{ij}^+} + \overline{c^- r_{ij}^-} , \\[2mm]
\overline{u_i''} &= C^+ C^- \overline{\varrho} \left(\frac{1}{\varrho^+} - \frac{1}{\varrho^-} \right) \delta U_i + \overline{c^+ u_i^+} + \overline{c^- u_i^-} , \\[2mm]
\overline{\varrho e u_i''} &= C^+ C^- \overline{\varrho} \left(\frac{\overline{c^+ \varrho e}}{\overline{c^+ \varrho}} - \frac{\overline{c^- \varrho e}}{\overline{c^- \varrho}} \right) \delta U_i + \overline{c^+ \varrho e u_i^+} + \overline{c^- \varrho e u_i^-} .
\end{aligned}
\tag{7.6}
$$

For the *advective* part of the turbulent flux of turbulent kinetic energy, the relation is more complicated, because it must be expressed as a function of the turbulent kinetic energies of the fluids within their own reference frames according to (3.12b). The demonstration given in D leads to:

$$\overline{\varrho k u_i''} = 3\,C^+ C^-\,\overline{\varrho}\left(\frac{\overline{c^+ \varrho k^+}}{\overline{c^+ \varrho}} - \frac{\overline{c^- \varrho k^-}}{\overline{c^- \varrho}}\right)\delta U_i + \overline{c^+ \varrho k^+ u_i^+} + \overline{c^- \varrho k^- u_i^-}$$
$$- (C^+ - C^-)\,C^+ C^-\,\overline{\varrho}\,[(\delta U)^2/2]\,\delta U_i\,.$$

(7.7)

Compared with the general expressions in (7.6), there is now an additional term *cubic* in δU, which is a directed flux of directed kinetic energy.

It must be emphasized that there is no explicit contribution of fluctuations in the turbulent concentration flux $\overline{\varrho c^\pm u_i''}$: by definition, the two-fluid approach entirely captures the flux of fluids by means of the evolution equations, instead of an algebraic closure of the turbulent diffusion as in the single-fluid case. *This is the fundamental reason for the superiority of the two-fluid approach for mixing instabilities.* Naturally, as already proposed [145], it is always possible to supplement a single-fluid statistical model with an evolution equation for the turbulent concentration flux. The two approaches become formally equivalent. However, here again, the directed and turbulent contributions in each term of the statistical equation of $\overline{\varrho c^\pm u_i''}$ are not separated, which makes modeling tricky.

It is of interest to compare the contributions of the various terms in the turbulent fluxes in (7.6) and (7.7):

- In the case of the Reynolds tensor, the "0D" approach of instabilities as described in Sect. 4.3 shows that the ratio of the absolute values of the directed and random contributions is of the order of K_D/K_B at the center of the mixing zone: typically a few percent in the absence of a driving term, but up to 25% for an RT instability. If we now compare the *components* of the Reynolds tensor along δU, direct simulations for the RT case show an even higher ratio, of the order of 50% [33].
- For the turbulent flux of turbulent energy, this ratio is often smaller: assuming that the turbulence levels are nearly the same in the two fluids, the linear directed contribution cancels, and the ratio of cubic directed and fluctuating contributions is thus of the order of $(K_D/K_B)^{3/2}$.[2]
- In the case of the turbulent mass flux, the situation is often reversed: in the limit of incompressible fluids, we have $c^\pm \varrho = \varrho^\pm$ (without the mean) and thus $\overline{c^\pm u_i^\pm} = \overline{c^\pm \varrho u_i^\pm}/\varrho^\pm = 0$ by the definition of $u^\pm$. Thus to within a factor, $\overline{u_i''}$ is identical to the turbulent concentration flux, $\overline{\varrho c^\pm u_i''}$.

---

[2] Note that the ratio behaves as $(K_D/K_B)^{1/2}$ if there is asymmetry between the turbulent fields.

- Finally, as already studied [139, 140], we note that for the turbulent flux of internal energy, the two-fluid expression is always compatible with changes of the thermodynamic references, in contrast to the single-fluid Boussinesq–Reynolds closure, for which it is necessary to impose $\sigma_e = \sigma_c$.

Relations (7.6) and (7.7) are rigorous, even for miscible fluids, and represent major realizability constraints for the single-fluid models. They are the basis for certain mixing flow models [34, 35] that could be described as degenerate two-fluid (or extended single-fluid) models. Furthermore, they provide insight on the nature of the "Rayleigh–Taylor" production term of the single-fluid approach. Indeed, according to (7.6), the single-fluid "Rayleigh–Taylor" production term of $\tilde{k}$ is written as:

$$-(\bar{p})_{,i}\,\overline{u_i''} = -C^+C^-\,\bar{\varrho}\,\left(\frac{1}{\varrho^+} - \frac{1}{\varrho^-}\right)(\bar{p})_{,i}\,\delta U_i - (\bar{p})_{,i}\,\left(\overline{c^+u_i^+} + \overline{c^-u_i^-}\right). \quad (7.8)$$

By comparison with (3.14), we then recognize the production terms of mean kinetic energies for each fluid, in $-(\bar{p})_{,i}U_i^{\pm}$, which must be present in order to feed the directed portion of the single-fluid turbulent kinetic energy, k_d, defined in (3.13). Moreover, as was noted above, the "Rayleigh–Taylor" terms for each fluid are not noticeable contributions to the production of $\tilde{k}$. Thus the intuitive interpretation in terms of the mean of the baroclinic term [142, §2.1] cited in Sect. 7.2, cannot be accepted. The two-fluid approach clearly identifies the turbulent energy production in (3.16) as the work of drag that dissipates the directed kinetic energy. Because k_d is not distinguished from $\tilde{k}$ in the single-fluid k–ε model, we cannot reconstruct an energy dissipation circuit equivalent to that of the two-fluid models.

7.4 The "Two-Fluid Extended" Single-Fluid Approach or "Drift Degenerate" Two-Fluid Approach

As was discussed above in Sects. 7.1 through 7.3, the single-fluid models cannot correctly capture RT instabilities because the models are too "simple"; in other words, they do not explicitly contain separate treatments for the turbulent and directed energies. Although these concepts are defined within the framework of a two-fluid analysis, they can still be incorporated into a single-fluid approach. The resulting descriptions can be equally called, depending on one's individual perspective, either extended single-fluid or degenerate two-fluid models.

One way of simplifying a two-fluid model to obtain a consistent single-fluid model is to degenerate it into a "drift" model, also known as a "drift flux" model. This involves replacing the two equations of fluid momentum, $\alpha^{\pm}\varrho^{\pm}U^{\pm}$, with one equation for the total momentum $\bar{\varrho}U$ and one algebraic closure of the velocity difference between the fluids, δU. The usual procedure for this, based on the example of the ASM (Algebraic Stress Model) [146], consists in writing the evolution equation of δU deduced from (3.9) with the closures of the model in Sect. 3.2:

$$\frac{\partial}{\partial t} \delta U_i + U_j^+ (U_i^+)_{,j} - U_j^- (U_i^-)_{,j}$$
$$= \frac{\partial}{\partial t} \delta U_i + (C^- U_j^+ + C^+ U_j^-)(\delta U_i)_{,j} + \delta U_j (U_i)_{,j} \pm (C^{\mp})_{,j} \delta U_j \delta U_i$$
$$= -\left(\frac{1}{\varrho^+} - \frac{1}{\varrho^-}\right) (\bar{p})_{,i} - \left(\frac{1}{\alpha^+ \varrho^+} + \frac{1}{\alpha^- \varrho^-}\right) D_i^* ,$$

(7.9)

and assuming that the time and space variations of δU are negligible. By also disregarding the coupling terms to the mean deformation (which is tensorial) and to the concentration gradient (quadratic), we obtain the relation:

$$D_i^* = \alpha^+ \alpha^- \frac{\varrho^+ - \varrho^-}{\bar{\varrho}} (\bar{p})_{,i} .$$

(7.10)

The expression for $\boldsymbol{D}^*$ from AWE's two-fluid model in (3.30) and (3.31) then leads to the algebraic closure:

$$\delta U_i = D_t \left(\frac{(\alpha^+ \varrho^+)_{,i}}{\alpha^+ \varrho^+} - \frac{(\alpha^- \varrho^-)_{,i}}{\alpha^- \varrho^-}\right) + \sqrt{\frac{\lambda_d}{C_d} \frac{|\varrho^+ - \varrho^-|}{\|\boldsymbol{\nabla} \bar{p}\|}} \, \text{Sgn} \, (\varrho^+ - \varrho^-) \frac{(\bar{p})_{,i}}{\bar{\varrho}} .$$

(7.11)

In addition to the turbulent dispersion term, δU contains a term in the direction of $(\bar{p})_{,i}$ which depends on the Atwood number and the characteristic size λ_d, but not on the volume fractions $\alpha^{\pm}$. Moreover, the contribution of the added mass disappears, because the temporal and spatial variations of δU are disregarded. Using this relation, the fluid velocities are substituted into all the equations of the model using the relation (3.8).

Here again, the behavior of AWE's "drift" two-fluid model can be comprehended using the "0D" approach. For this, there is no need to repeat all the explanations given for the original model in Sect. 6.1, because the "drift" closure provided here for the momentum transposes in "0D" onto the directed energy. Thus we need only return to the equations of the complete model (6.5) and (6.10) where $dK_D/dt = (dL/Ldt)K_D = 0$ is imposed, and eliminate the contribution of the added mass in the expression for Π_D in (6.6). The resulting equations are solved along the same lines as for the complete model, keeping condition (6.8). The results are summarized in Table 7.2.

A comparison of the results for the complete and "drift" versions of AWE's two-fluid model show discrepancies of the order of 20% for the growth rates and K_B/K_D ratios, while the other parameters are fairly close. Thus within this error margin, the behaviors as a function of C_i and C_d are entirely comparable (see Figs. 6.1 and 6.2, p. 71 and 72; the equivalent for the drift model will not be given here). We also note the complete decoupling of all the quantities from coefficient C_d in the KH and RM cases.

Table 7.2. Basic geometric and energy characteristics of KH, RT, and RM turbulent mixing zones as reconstructed by the "drift" version of AWE's "0D" simplified two-fluid model. The boxed numerical values were obtained with the standard coefficients of AWE's two-fluid model given in Table 3.1, p. 27. All the results are very close to those of the standard model in Table 6.1, p. 68

Instability	KH	RT	RM
Constant	$\mathcal{X}_A = \dfrac{4C_i^2}{\sqrt{4C_i^2 + C_\mu}}$ $\boxed{\approx .12}$	$\mathcal{Y}_A = \dfrac{1}{C_d\left(1 - \sqrt{2}C_i\,\mathcal{U}_{\mathrm{RT}}\right)^2}$ $\boxed{\approx .123}$	$n_A = \dfrac{8C_i^2}{12C_i^2 + C_\mu}$ $\boxed{\approx .4}$
L	$\mathcal{X}_A \times \Delta U_y t$	$\mathcal{Y}_A \times A\Gamma t^2$	$L_0\left(\dfrac{t}{t_0}\right)^{n_A}$
K_I	$\dfrac{1}{12} \times \Delta U_y^2$	$\dfrac{\mathcal{Y}_A}{12} \times (A\Gamma t)^2$	$K_{I0}\left(\dfrac{t}{t_0}\right)^{-n_A}$
K_D	$\dfrac{2}{\zeta} \times \dfrac{\mathcal{X}_A^2}{48} \times \Delta U_y^2$	$\dfrac{\mathcal{Y}_A^2}{12} \times (A\Gamma t)^2$	$\dfrac{n_A^2}{48} \times \left(\dfrac{L_0}{t_0}\right)^2 \left(\dfrac{t}{t_0}\right)^{-2(1-n_A)}$

Table 7.2. continued

Instability	KH	RT	RM
		$\mathcal{Y}_A\,\mathcal{U}_{RT}^2(C_\mu, C_d, C_i)$	
K_B/K_I	$\dfrac{2C_i}{\sqrt{4C_i^2 + C_\mu}}$ $\boxed{\approx .33}$	$\boxed{\approx .73}$	
		$\mathcal{U}_{RT}^2(C_\mu, C_d, C_i)$	
K_B/K_D	$\dfrac{\zeta}{2C_i^2}$ $\boxed{\approx 45.4}$	$\boxed{\approx 5.94}$	$\dfrac{1}{2C_i^2}$ $\boxed{\approx 45.4}$
κ_b	$\dfrac{C_i}{\sqrt{2}\,C_\mu}$ $\boxed{\approx .82}$	$\dfrac{C_i}{2C_\mu}$ $\boxed{\approx .58}$	

$$\sqrt{2}C_\mu\,\mathcal{U}_{RT}^3 - 8C_i(2C_dC_i^2 - 1)\mathcal{U}_{RT}^2 + 16\sqrt{2}C_dC_i^2\,\mathcal{U}_{RT} - 8C_dC_i = 0\ .$$

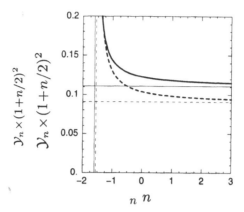

Fig. 7.1. Evolution, as a function of the exponent n, of the normalized growth coefficient of the mixing zone, $\mathcal{Y}_n \times (n+1)(n+2)/2$, for a self-similar Rayleigh–Taylor instability with acceleration varying as t^n, as obtained using the degenerate "drift" version of AWE's two-fluid model (*solid line*). For comparison, the result for AWE's complete two-fluid model is also shown (*dashed line*). The asymptotes are also shown (*thin lines*)

The response to an acceleration varying as t^n is calculated in the same way as for the complete model, and yields:[3]

$$
\begin{cases}
L & = \mathcal{Y}_n \times \mathcal{A}\Gamma t^2 = \dfrac{4/(2+n)^2}{C_d \left(1-\sqrt{2}C_i\,\mathcal{U}_n\right)^2}\,\mathcal{A}\Gamma t^2, \\[2ex]
K_I & = \dfrac{(1+n/2)\mathcal{Y}_n}{12\,(1+3n/4)} \times (\mathcal{A}\Gamma t)^2, \\[2ex]
K_D & = \dfrac{(1+n/2)^2\mathcal{Y}_n^2}{12} \times (\mathcal{A}\Gamma t)^2, \\[2ex]
K_B & = \mathcal{U}_n^2(c_\mu, c_d, c_i) \times K_D, \\[2ex]
\kappa_b & = \dfrac{C_i}{2C_\mu},
\end{cases}
\tag{7.12}
$$

with the following equation in $\mathcal{U}_n$:

$$
\sqrt{2}C_\mu\,\mathcal{U}_n^3 - 8C_i\left[2C_dC_i^2 - 2\frac{1+3n/4}{2+n}\right]\mathcal{U}_n^2 + 16\sqrt{2}C_dC_i^2\,\mathcal{U}_n - 8C_dC_i = 0,
\tag{7.13}
$$

As with the complete model, all the quantities remain finite and reasonable for all $n > -1$, and converge toward the asymptotic limits:

$$
\mathcal{Y}_\infty \approx .111 \times (1+n/2)^2 \quad \text{and} \quad (K_B/K_D)_\infty \approx 4.91,
\tag{7.14}
$$

---

[3] Ⓤ New developments on SSVARTs [54,55] discussed in Sect. 9.3 have noticeably *changed* and sometimes *contradicted* the findings in the remainder of this part.

which can be compared with (6.14). As shown in Fig. 7.1, the relative discrepancy between the two versions is fairly constant and is of the order of 20% as $n \to \infty$, which might seem surprising since, a priori, the "drift" model should not capture large transient effects very well. Below $n = -1$, the "drift" model diverges for $n = n_0 \approx -1.60$, a value very close to that observed for the complete model.

Therefore the "drift" model appears as an interesting compromise between complexity and performance, because the observed discrepancies can be readily compensated by calibrating the coefficients (which will not be elaborated here). Moreover, a single-fluid model can be readily obtained by degenerating the internal energy equations. This would make it possible to extend the closures to render the single-fluid models compatible with the two-fluid nature of the flow. The identities (7.6) combined with (7.11) give an example of what such an "extended two-fluid" k-ε model *might* contain:

$$
\overline{\varrho c^{\pm} u_i''}^{\mathrm{m}} = -\overline{\varrho}\, D_t \left(C^{\pm} \right)_{,i} + C^{+}C^{-}\sqrt{\frac{\lambda_d}{C_d}\frac{|\varrho^{+}-\varrho^{-}|}{\|\nabla \overline{p}\|}}\;\mathrm{Sgn}\left(\varrho^{+}-\varrho^{-}\right)(\overline{p})_{,i}\,,
$$

$$
\overline{r_{ij}} \stackrel{\mathrm{m}}{=} \frac{1}{C^{+}C^{-}\,\overline{\varrho}}\;\overline{\varrho c^{\pm} u_i''}\,\overline{\varrho c^{\pm} u_j''} + {}^{2}\!/\!_{3}\,\tilde{k}\,\delta_{ij} - \nu_t\,\overline{\varrho}\left[(U_i)_{,j} + (U_j)_{,i}\right],
$$

$$
\overline{u_i''} \stackrel{\mathrm{m}}{=} \left(\frac{1}{\varrho^{+}} - \frac{1}{\varrho^{-}} \right)\overline{\varrho c^{\pm} u_i''}\,.
$$

(7.15)

We now see complementary terms on the Boussinesq–Reynolds closure, which somewhat resemble those obtained by the ASM method [146]. An approach of this type has already been the subject of theoretical and numerical studies by A.V. Polyonov at the VNIITF in Russia under the name "heterogeneous k-ε" or "Hk-ε" [18 21], but it does not seem to have been used in a production code.

The "drift" approach is the basis for a simple approximation of the RT mixing zone growth in the case of a randomly variable acceleration [31,32,118, 119], which is in fairly good agreement with experimental observations [118–121]. According to (7.11), neglecting the turbulent diffusion and assuming a constant ratio of λ_d over L, we obtain the relation $\delta U \propto L' \propto \sqrt{L \mathcal{A}\Gamma}$, which when integrated becomes:

$$
L(t) \approx \mathcal{Y}_0 \mathcal{A} \left(\int_0^t \sqrt{\Gamma(s)}\, \mathrm{d}s \right)^2.
$$

(7.16)

The normalization in $1/(1+n/2)^2$ used in the graphs of $\mathcal{Y}_n$ in Figs. 6.3 and 7.1 was obtained by applying this formula to $\Gamma(t) \propto t^n$. Thus it is not surprising that it makes coefficient $\mathcal{Y}_n$ nearly constant, in contrast to the free fall behavior in $1/(1+n)(1+n/2)$. Approximation (7.16), which derives from the formulation of drag in a Newtonian regime, is thus characteristic of the two-fluid nature of RT flows.[4]

---

[4] Ⓤ Equation (7.16) seem's to have been first derived on somewhat different arguments by K.I. Read [118], hence its designation as "Read's formula" in later publications [54, 55].

Specific Treatment of Shocks: Constraints on Models[1]

8.1 Qualitative Phenomenology of Interaction between a Shock and Velocity or Density Heterogeneities

Before examining the phenomenology of the interaction between a shock and turbulent or material heterogeneities, it is useful to recall that the turbulence of a compressible fluid has *two components* that are very different in nature. As is reviewed in B, the velocity fluctuation field can be decomposed into divergent irrotational and nondivergent rotational parts, respectively [147–149]. Only the latter is present for an incompressible fluid. Despite certain formal difficulties and the fact that the two fields evolve in a coupled way, this decomposition provides an intuitive understanding of the nature of compressible turbulence and the effect of shock waves [148, 149]:

- The rotational part is associated with the vorticity and is thus essentially transported by random eddies in what appears on average as a diffusion-like process. It is dissipated by the inertial cascade down to the Kolmogorov scale.
- In contrast, the divergent part is associated to exchanges with the internal energy of the medium by means of volume variations and thus experiences a wave like propagation at a velocity similar to that of sound. Its dissipation is dominated by disordered "shock wavelets" induced by nonlinear

---

[1] One could be surprised by the notably different lengths of the discussions devoted to self-similar incompressible instabilities on one hand (Chaps. 4 through 7) and shocks on the other hand (Chap. 8), given the importance of the latter for the DAM. This reflects how greatly the literature on these two subjects differs with regard to both phenomenological understanding and sophistication of models. Therefore, this chapter is primarily a qualitative discussion of a research strategy, rather than a quantitative and critical review of existing models.

Ⓤ Since the first version of this chapter was written in early 1999, little if no progress has been recorded in the literature. The modeling strategy discussed here stands today just as it did then.

A. Llor: *Statistical Hydrodynamic Models for Developed Mixing Instability Flows*, Lect. Notes Phys. **681**, 93 106 (2005)
www.springerlink.com

Fig. 8.1. Schematic representation of the interaction of a shock with a turbulent medium (numbering same as in text): (*0*) initial state of isentropic, isotropic, homogeneous, subsonic compressible turbulence; (*1*) main shock zone; (*2*) zone of supersonic turbulence, of negligible thickness in the limit of weak shocks; (*3*) zone of subsonic turbulence and strong compression; (*4*) zone of anisentropic, anisotropic, subsonic compressible turbulence (slow relaxation). The usual turbulence models based on the assumption of weak gradients are only applicable to zones 0 and 4. The thickness of zones 1, 2, and 3 is often negligible, so the shock description can be reduced to the jump relations between states 0 and 4. The phenomenology is equivalent for the interaction with a heterogeneous medium

interactions. This dissipation is small when the turbulent Mach number, $\tilde{k}^{1/2}/c_s$, is less than one, but becomes very efficient in the transonic region, to the extent that supersonic isotropic homogeneous turbulence has never been observed (nor treated theoretically) [148, 149].

Shock passage through a turbulent medium is a problem of greatest interest in aerodynamics, and has thus been extensively studied since the 1950s [60, 150, 151]. However, quantitatively relevant direct 3D simulations of the simplest cases, such as a plane shock in a gas with isotropic homogeneous turbulence [152–157], have become available less than ten years ago. As a result, many of the uncertainties regarding experiment interpretations have been resolved, and a basic qualitative description is now available that we shall here extend to the case of a heterogeneous fluid [60]. Globally, the interaction with a turbulent medium causes the physical shock to spread over a finite width, within which a succession of four very different dissipative zones can be distinguished [60, 152–157], as diagrammed in Fig. 8.1:

1. **Main shock zone, or near field zone:** the initially plane shock surface is distorted through interactions with heterogeneities, but in the steady regime it remains, on average, plane with a uniform velocity if the initial medium is, on average, homogeneous. For a strong or medium-strength shock, the major portion of the dissipation occurs as the wave passes, perturbing the characteristic heterogeneities of the medium through two distinct mechanisms:

- by creating entropy heterogeneities due to velocity fluctuations of the medium (which may be isentropic in the upstream state); these are especially noticeable for shocks at Mach greater than 1.5;
- by creating velocity fluctuations due to heterogeneities of the thermodynamic state of the medium (which may be nonturbulent in the upstream state): these are more intense as the Atwood number departs from zero.

The resulting turbulent velocity field downstream is not in spectral equilibrium and displays:

- an increase of divergent fluctuations (which can become supersonic if the shock is sufficiently strong);
- practically unchanged longitudinal fluctuations of vorticity;
- an increase of transverse fluctuations of vorticity due to the decrease in the gradient lengths in the direction of the shock;
- an increase in the microscopic dissipation rate $\bar{\varepsilon}$,
- in the two-fluid case, an increase in the transverse vorticity fluctuations and the interpenetration velocity at small scales, due to the differences in shock velocities and impedances between the two media;
- and last, a decrease of most of the characteristic fluctuation lengths (integral, Taylor, Kolmogorov, fluid dispersion, and other scales).

2. **Zone of secondary shocks and evanescent waves:** the secondary shock wavelets created by the multiple reflections, refractions, and relaxations of the principal shock against heterogeneities are rapidly dissipated, bringing turbulence down to a subsonic regime, however without restoring spectral and isentropic quasi-equilibria; at the same time, the nonpropagating or evanescent acoustic modes relax. The thickness of this zone becomes negligible in the limit of weak shocks.

3. **Zone of redistribution of turbulent modes:** because of nonlinearities of the hydrodynamic equations, the excess divergent turbulence is transformed into rotational turbulence or is transported far downstream by acoustic waves, and the correlations between vorticity and the principal directions of deformation return to normal.

4. **Zone of relaxation toward spectral equilibrium, isotropy, and isentropy, or far-field zone:** of these three slow mechanisms dominated by turbulent dissipation, the first two are included in most models, but the last is often unduly ignored. Shock–turbulence interaction creates heterogeneities of the stagnation thermodynamic state, even if it is uniform in the upstream state, and its further relaxation is due to turbulence.

The usual turbulence models are applicable only in the *last* of these four zones, where the gradient lengths of the various mean quantities are not shorter than the characteristic turbulent scales. Modeling of the first three zones involves specific problems, which might require the introduction of additional variables and statistical correlations such as divergent and rotational fractions of the turbulent fluctuations, non-normality of correlations, anisotropy, etc.

However, assuming that the thickness of these three zones is negligible compared with the scales of the investigated flow, or with the characteristic thickness of the numerical spread of the shocks, then the description of these zones can be simply reduced to giving the *jump relations* for the otherwise already modeled quantities.[2]

The interaction between shock and heterogeneities has the same global phenomenology as shock–turbulence interaction, with several adaptations. The rapid relaxation behavior of the divergent component of the turbulent velocity field is conserved in zones 2 and 3. By contrast, the possibly large discrepancy between the Hugoniot curves of the system components will induce large differences in the *mean* velocities of the different heterogeneities at the beginning of zone 2. In the two-fluid approximation, if $[\![U^{\pm}]\!]$ are the velocity jumps of the two components as given by their respective Hugoniot curves for a common pressure jump equal to the jump of the mean shock pressure, $[\![\overline{p}]\!]$, the difference between mean velocities in zone 2 will be estimated as $\delta U \approx [\![U^{+}]\!] - [\![U^{-}]\!]$. Depending on the characteristic dimensions of the heterogeneities perturbed by the shock passage, the velocity differences will then relax more or less quickly due to drag, by converting a portion of the mean directed kinetic energy into turbulent kinetic energy. The modeling of drag, which is crucial at this level, cannot be reduced to usual expressions of type (3.30), which are valid only in the far field.

8.2 Preliminary Comments on Modeling of Shock–Heterogeneity Interaction

For industrial applications, modeling of shock–heterogeneity interaction is basically reduced to the shock–turbulence case, and with widest use in aerodynamics. Thus most models applied to this problem are used in a particular context: steady flows, shock-induced energy dissipation into mostly internal energy, dominant effect of boundary layer in production and destruction of turbulence, emphasis on the reconstruction of the Reynolds tensor (drag) and thermal flux (heating), etc. The adaptation of these models to capture shocks

---

[2] This analysis might seem incorrect in the limit of weak shocks, at $[\![U^{2}]\!] \ll \tilde{k}$, where the turbulent viscosity could, in place of the molecular viscosity, be the dissipative mechanism to stabilize the shock front. If this mechanism was dominant, the spreading thickness of the shock would be of the order of $\lambda_{c} \approx \tilde{k}^{2}/(\tilde{\varepsilon}\, c_{s})$ (where c_{s} is the mean sound velocity). In fact, the turbulent viscosity is a relevant quantity only at dimensions greater than the integral length scale given by (2.15), and the condition $\lambda_{i} \ll \lambda_{c}$ then leads to $c_{s}^{2} \ll \tilde{k}$, that is, a supersonic turbulence! Thus in all practical cases, the molecular-scale dissipation mechanisms do stabilize the compression waves. Although the hydrodynamic limit is not strictly valid at these scales, these mechanisms are reasonably represented by the molecular viscosity and the molecular thermal conduction (associated, respectively, with τ_{ij} and θ_{i} in (2.1), and with $\overline{\tau_{ij}}$ and $\overline{\theta_{i}}$ in (2.6) and (2.11)).

is often rudimentary: the energy in the shock zone is dissipated into internal energy by artificial viscosity or by a Riemann solver, and limiters are applied to the first gradient closures to avoid diverging quantities in the shocks.

This approach appears to be acceptable (and accepted) in many aerodynamic applications. However, as was generally discussed in Sect. 8.1 and will be demonstrated using an example in Sect. 8.4, it cannot describe, even approximately, the evolution of the major turbulent quantities. Considering the importance of turbulence in the evolution of mixing layers, the DAM applications require a more refined modeling of shocks. We will use the approach mentioned in Sect. 8.1, where only the *jump relationships between the upstream and downstream zones far from the shock* are described (zones 0 and 4 in Fig. 8.1).

Within this approach, there are major difficulties in physically describing shocks in turbulent and/or mixing environments. Indeed, and contrary to what is often claimed [158], *jump relations cannot be obtained merely by considering weak solutions (in the distributions' sense) of equations verified in the absence of shocks* [159]:

1. A shock always involves a mechanism that dissipates kinetic energy. In a simple non-turbulent fluid, this dissipation must lead to internal energy. For a given shock velocity, the conservation equations of mass, momentum, and energy are then sufficient to uniquely establish the downstream characteristics (whence the Hugoniot curve), and this appears as *completely independent of the dissipation mechanism* which is specific to the fluid being considered [160, 161]. However, in a turbulent fluid, a heterogeneous mixture of fluids, or a fluid lacking local thermodynamic equilibrium, there are several reservoirs of energy that can absorb dissipation: turbulent kinetic energy, internal energies of the various fluids, ionic, electronic, and radiative energies, etc. The Hugoniot cannot be determined based on the conservation relations alone, because the distribution of dissipation is related to the *particular nature of the dissipative mechanisms in the system* [159–161]. Thus a suitable treatment of shocks requires explicit hypotheses for these mechanisms, as well as a closure in the statistical models.[3] In the particular case of shock–turbulence interaction, these hypotheses are not crucial for stability, because the internal energy most often dominates the turbulent kinetic energy, and the dissipation in a shock essentially goes into internal energy (even though the jump of $\tilde{k}$ must be specifically described, see Sect. 8.4).

2. In the two-fluid case, the above problem is exacerbated by the large number of energy reservoirs, and especially by the possibility of *momentum exchange*. For example, in AWE's two-fluid model, when the drag and Reynolds stresses are disregarded, integrating the momentum equations

---

[3] It is even imaginable that there can be shock separation phenomena, like those observed in radiative hydrodynamics or for materials undergoing phase changes [160, 161].

over a stationary discontinuity along the x axis yields the jump relations:

$$[\![\alpha^{\pm}(\varrho^{\pm}U_x^{\pm}U_x^{\pm} + P)]\!]_0^1 = \int_0^1 (\alpha^{\pm})_{,x} \, P \, dx = \pm \mathcal{F} \, . \qquad (8.1)$$

The right hand side cannot be integrated unambiguously in order to fully define the jump relation, because it is not the divergence of a flux. This is a momentum exchange term within the discontinuity which, assuming that P and $\alpha^{\pm}$ are monotonic, can take *any value* between $P_0[\![\alpha^{\pm}]\!]_0^1$ and $P_1[\![\alpha^{\pm}]\!]_0^1$. In fact, the zero discontinuity is a limit that does not need to be considered in the case of multifluid flows, even though it retains a physical sense in the single-fluid laminar case. Indeed, as was presented in Sect. 8.1 and illustrated in Fig. 8.1, a shock interacting with heterogeneities and *described as an ensemble mean*, is necessarily *spread* and involves *redistribution* mechanisms between the various contributions to momentum and energy. The exchange term $\mathcal{F}$ in (8.1) thus characterizes the global balance of these redistribution mechanisms, and only one jump relationship for the *total* momentum is available.

3. In the two-fluid case, shocks induce an increase in the turbulent kinetic energy by interacting both with the turbulent velocity field and with the heterogeneous mixed structure. These two distinct mechanisms (each is observable in the absence of the other) thus must be modeled.

4. Besides the energy aspect, a shock generates perturbations of all the parameters describing the flux terms and conversion terms in the conservation equations. In particular, the major propagation direction of the shock necessarily changes the anisotropy characteristics of the fluctuations (for example, for the Reynolds tensor). The other important quantity for describing turbulence is the dissipation rate of turbulent kinetic energy, $\tilde{\varepsilon} = -\overline{\tau_{ij}(u_i'')_{,j}}/\overline{\varrho}$, whose evolution equation is modeled heuristically from the equation in $\overline{\varrho k}$ without taking into account the effect of possible shocks. All these quantities must therefore be specifically modeled in the presence of a shock.

Along with these difficulties related to the description of dissipative phenomena in a shock, there are also difficulties associated with possible spurious dissipations introduced by the physical modeling and numerical treatment:

5. At the modeling level, Boussinesq–Reynolds algebraic gradient closures yield divergent and non-physical turbulent fluxes in the shock zones, because the characteristic gradient lengths are much smaller than the turbulence scales (see Sect. 8.1). In the absence of other modelings, it is thus mandatory that all closures that might potentially display such behavior be subjected to adiabatic corrections (discussed in E) and limiters, for example those deduced from realizability conditions [61, §4.6], [139, 140]. In the case of weak shocks, which are by definition quasi-isentropic, adiabatic correction of the fluxes can entirely correct the nonrealizability.

6. At the numerical level, an artificial dissipation mechanism, usually viscous, is introduced to generate the entropy source during shock passage [81]. Its value, which is much greater than that of the physical dissipative phenomena. is adjusted so that the shock zone typically spreads over two or three calculation mesh cells. As has been discussed elsewhere for a simple non-turbulent fluid [162], the detection of shock zones and the calculation of artificial viscosity stresses must always be deduced from an analysis of the tensor of the velocity field gradients, for reasons of thermodynamic consistency that will not be reexamined here. In simplified terms, it is assumed that shocks are present only at points where the velocity field, filtered of any possible isentropic compressions, displays a direction of fluid compression. In the case of compressible turbulence, the Favre averaged velocity field depends partly on the dynamics of the density fluctuations, and thus does not allow one to *specifically* characterize the fluid compression, as the expression for div U in (7.2) shows. An artificial dissipation mechanism must thus introduce the *Reynolds averaged* velocity field. We emphasize that giving both the Reynolds and Favre averages for u determines the turbulent mass flux, $\overline{u''}$, which appears in the "Rayleigh–Taylor term" and the modeling of which was discussed in Sect. 7.2.

7. The retained modeling must not depend on the characteristic numerical spread length of shocks as defined by the artificial viscosity coefficients. Indeed. the treatment of shocks by an artificial dissipative term is only permitted to the extent that the downstream and upstream states are independent of the form and value of this term. Therefore, Boussinesq–Reynolds gradient closures, which do comply with this constraint in their natural formulation (2.16), can pose a problem whenever they are subjected to a limiter. because the fluxes are no longer dependent on the gradients. Solving simultaneously this point and item 5 above can be somewhat difficult.

Overall, neither the single- nor two-fluid approaches to the mixing flow modeling provide any particular simplifications for these problems, except for item 6, where the two-fluid description is clearly superior and will be discussed below.

8.3 On Artificial Dissipation in a Two-Fluid Medium

As was discussed in Sects. 8.1 and 8.2, the modeling of the interaction between a shock and a two-fluid medium must include the details of the dissipative mechanisms, because a weak solution of the conservation equations cannot yield the jump relations (between zones 0 and 4 in Fig. 8.1). Two approaches are possible:

- either the precise redistribution mechanism of the modeled quantities is known (in zones 1 through 3 in Fig. 8.1) and it can be inserted into the equations at the cost of likely complications (such as taking into account the divergent portion of the turbulent kinetic energy);

- or the precise mechanism is not known except for its *global effect* (between zones 0 and 4 in Fig. 8.1) and it suffices to supplement the system equations with corrective terms that are totally arbitrary but chosen so as to *reconstruct the same global behavior.*

The introduction of artificial terms to reconstruct the *physics* of the jump relations in a shock is reminiscent of the *numerical* technique of artificial viscosity for capturing shocks [81, 162]. Therefore, the artificial terms for the physical reconstruction of jumps also represent a numerical artificial dissipation mechanism, provided that they are sufficiently strong to spread the shocks over a thickness greater than the calculation mesh cell. Stretching the terminology, these physical terms will also be called artificial dissipations.

For reasons of simplicity and considering the embryonic state of two-fluid studies at the DAM, we shall consider below only the artificial dissipative solution, for which we shall examine the *minimal* formulation in the case of a generic two-fluid model. We shall take up here the so-called "six-equation" two-fluid model, to which AWE's two-fluid model reduces by eliminating all but the conservation equations of mass, momentum, and internal energy.

In a purely heuristic manner, we can first establish an inventory of *all* the allowable conservative dissipation mechanisms in a two-fluid system:

- two viscosities (one per fluid),
- two thermal conductions (one per fluid),
- one drag (exchange of momentum between fluids),
- and one thermal exchange between fluids.

Of these six mechanisms, we need retain only three: one basic dissipation mechanism to capture the shock, and two supplementary mechanisms to model the exchanges of momentum and energy during the shocks. Since a conductive mechanism alone cannot stabilize strong shocks in the single-fluid case [160–162], we will retain below only the three mechanisms of viscosities and drag as artificial dissipative terms.

Let us examine the case of scalar artificial viscosity stresses, $Q^\pm$ (the tensorial extension does not pose any problems), and, for the time being, in the absence of artificial drag. The fluid compressions are related to the Reynolds averaged compression, $(V_i)_{,i}$, and then the artificial viscosity terms to be added to the internal energy equations are:

$$-\beta^\pm Q^\pm (V_i)_{,i} \quad (\geq 0) . \qquad (8.2)$$

The factors $\beta^\pm$, which might have been grouped with the $Q^\pm$, were conserved by analogy with the terms of the work of pressure in (3.23). By conservation of energy, the sum of the artificial viscosity terms in the kinetic energy equations is then:

$$-(\beta^+ Q^+ + \beta^- Q^-)_{,i} V_i = -(Q)_{,i}(\alpha^+ U_i^+ + \alpha^- U_i^-) , \qquad (8.3)$$

from which we obtain the artificial viscosity terms to be added to each momentum equation:

$$-\alpha^{\pm}(Q)_{,i} \, . \tag{8.4}$$

This modeling can be compared with that of AWE's in Sect. 3.2 where "two pressures" are reduced to just one in the equations of $U^{\pm}$. It has been successfully used in AWE's two-fluid model whenever the more simplistic formulation with only one artificial viscosity described in Sect. 3.2 caused instabilities [91].

The artificial viscosities for each fluid can be expressed classically as combinations of linear and quadratic functions of the compressions in each fluid:

$$Q^{\pm} = -\, a_1 \, \varrho^{\pm} c_s^{\pm} \left(\frac{\beta^{\pm}}{\alpha^{\pm}} \, \min(0, V_{i,i}) \ell \right) + a_2^2 \, \varrho^{\pm} \left(\frac{\beta^{\pm}}{\alpha^{\pm}} \, \min(0, V_{i,i}) \ell \right)^2 \, , \tag{8.5}$$

where ℓ is the characteristic cell size, and a_1 and a_2 are dimensionless adjustment coefficients (this formulation is not that of AWE's two-fluid model) [91]. We note that the total artificial viscosity $Q = \beta^+ Q^+ + \beta^- Q^-$ can no longer be expressed as a function only of the global compression, $(V_i)_{,i}$, and the average density and sound velocity. Therefore, this approach can do well in distinguishing, although not necessarily in a precise manner, the four types of two-fluid shocks, namely, weak or strong on one or the other fluid.

The addition of artificial drag to the two artificial viscosities is less trivial for three reasons:

• The energy transfer associated with the artificial drag presumably does not involve turbulent kinetic energy (as does ordinary drag), but rather internal energy: the exchanges with *nondivergent rotational* turbulence occur over a characteristic time much longer than that of the shock;
• the passage of a shock through a two-fluid medium can result, depending on the case, in an increase or decrease in the directed kinetic energy, and thus the sign of the artificial drag is indeterminate: it does not necessarily represent a dissipative mechanism;
• as a complement to the previous point, we note that there are now only *two* thermodynamic constraints on the entropies of the fluids, for *three* artificial dissipative mechanisms: the constraints on the $Q^{\pm}$ in (8.2) might be narrowed or relaxed as a result.

For these reasons, we shall not attempt to give a precise formulation of the three artificial dissipative terms combined. Eventually, such a formulation would have to, first, rely on more precise and material-dependent shock data and, next, be adapted to the specific features of the numerical scheme being used, particularly regarding the treatment of the exchange terms $(\alpha^{\pm})_{,x}P$ in (8.1). Therefore, at this stage we shall merely review the global structure of these terms, to be added to the right-hand side of the two-fluid energy balances in (3.14):

$$\begin{array}{c} \frac{\partial}{\partial t}\overline{c^{\pm}\varrho e}\ldots\ldots \\ \frac{\partial}{\partial t}\overline{c^{\pm}\varrho k}\ldots\ldots \\ \frac{\partial}{\partial t}\overline{c^{\pm}\varrho U^2/_2}\ldots \\ \frac{\partial}{\partial t}\overline{c^{\pm}\varrho f}\ldots\ldots \end{array} \left[\begin{array}{ccc} . & -\beta^{\pm}\,Q^{\pm}\,(V_i)_{,i} + \omega^{\pm}\,F_i\,\delta U_i & \boxed{\geq 0} \\ . & . & . \\ -\alpha^{\pm}\,(Q)_{,i}\,U_i^{\pm} & . & \mp\quad F_i\,U_i^{\pm} \\ -\alpha^{\pm}\,(Q)_{,i}\,U_i^{\pm} - \beta^{\pm}\,Q^{\pm}\,(V_i)_{,i} & . \end{array} \right],$$

$$(8.6)$$

with the condition on the distribution coefficients $\omega^+ + \omega^- = 1$, and the two thermodynamic constraints marked by the boxed inequalities shown above.

At this point we must emphasize the numerical importance of the artificial dissipative approach. Spatial discretization in a finite volumes numerical scheme always produces discontinuities at the nodes, branches, or faces of the mesh cells. Godunov-type hydrodynamic codes solve the Riemann problem for these numerical jumps. The physical jumps, which are the shocks, are then represented and captured by a few successive numerical discontinuities and are thus "automatically" treated. In the single-fluid laminar case, this poses no problem, because the jump relations are independent of the details of the dissipative mechanisms [81, 162], but in a two-fluid medium, the treatment of shocks using a Godunov-type scheme is then fixed by the more or less arbitrary choices realized in the Riemann solver, in particular for the exchange terms [163, §II.5]. In contrast, schemes with explicit artificial viscosity, such as the single-fluid VNR [81] or the two-fluid "two artificial viscosities and one artificial drag" given above, have greater flexibility in modeling shocks. At any rate, in all the two-fluid schemes, the perturbations due to numerical discontinuities must be minimized. Indeed, in the single-fluid laminar case, the only perceptible effect due to numerical discontinuities is an excess dissipation in isentropic evolution zones, which is usually weak and uniform. By contrast, in the two-fluid case, exchange terms could induce an erratic distribution of the dissipation: without violating the second law of thermodynamics, the dissipation could be concentrated, for example, in only one of the fluids (always, or in every other mesh, etc.) Fortunately, the presence of dissipative exchange terms in the physical models often has a regularizing influence.

8.4 Basic Bias Estimates of DAM's k–ε and AWE's Two-Fluid Models Applied to Shock–Turbulence Interaction

The production of turbulent kinetic energy due to a shock was not considered in Sect. 8.3 above because:

- it is marginal compared with the production of internal energy and thus does not perceptibly change the shock propagation conditions, and
- in the two-fluid case, it is often dominated by the conversion of directed kinetic energy into turbulence downstream of the shock due to the effect of the work of drag.

Thus it is presumably more important to have a good modeling of the production and relaxation of directed kinetic energy before and after the shock, respectively, but strictly speaking, it would be necessary in (8.6) to divert a portion of the dissipation into the turbulent fields. This dissipative coupling becomes necessary if we want to correctly model the production of $\tilde{k}$ in the single-fluid case of shock–turbulence interaction. The response of DAM's single-fluid k-ε and AWE's two-fluid models in this situation will provide an example.

The case of the interaction of a shock with a turbulent mixing zone is discussed in [142], with a particular analysis of the influence of the closure of the "Rayleigh–Taylor" production term of DAM's k-ε model. This situation is complicated to analyze, and numerical experiments are needed to evaluate the influence of the various effects: type of closure (adiabatic corrections, higher-order models, see Chap. 9), mesh refinement, existence of a counter-gradient zone within the numerical spread zone of the shock, direction of the shock relative to the stratification, etc.

We shall limit the discussion to the simpler and analytic situation of the interaction between a plane normal steady shock and a homogeneous isotropic turbulence. A numerical study of DAM's k-ε model applied to shock–turbulence interaction has already been made in order to interpret an experiment [26, §4.1], with emphasis on the relaxation after the shock. Here we will concentrate on the aspect of the production of $\tilde{k}$ and $\tilde{\varepsilon}$ by the shock. In this simple case of shock–turbulence interaction, we note that the "Rayleigh–Taylor" production term, as closed and limited in (2.17), cancels because the mean density and pressure gradients are not opposed (without the limitation the situation would be even worse, with a destruction of turbulence).

AWE's two-fluid model cannot describe single-fluid turbulence in a strict sense: the integral length scale, λ_l, which is required to close fluxes, is defined only in mixing zones in relationship with the drag scale λ_d. To evaluate the behavior in our chosen ideal situation, we must assume that the turbulent medium is actually two-fluid, with the two components having identical characteristics. Thus the fluids behave as passive scalars, and their volume fractions are irrelevant; only λ_d must be uniform in the upstream medium.

The analysis is simplified by using a "0D" approach. A steady shock propagating in the x direction is represented as a narrow ditch in the axial deformation rate $(U_x)_{,x}$ (all other components cancel) and we use a Lagrangian description: in other words, $(U_x)_{,x}$ is applied to each fluid particle during time:

$$\Delta t = \frac{-1}{(U_x)_{,x}} \text{Log}\left(\frac{\varrho_1}{\varrho_0}\right) , \qquad (8.7)$$

where ϱ_1/ϱ_0 is the total compression ratio between upstream and downstream. The "ideal" shock is thus obtained in the limit $\Delta t \to 0$ with $\Delta t (U_x)_{,x}$ constant. Following the argument presented above in Sects. 8.1 and 8.2, the integration of the model equations at the time of shock passage should depend on Δt and on the detailed profile of $(U_x)_{,x}$. In fact, these equations have invariants in the limit $\Delta t \to 0$, which makes it possible to find jump relations.

When $\Delta t \ll \tilde{k}/\tilde{\varepsilon}$, the dissipation and flux terms are negligible and, in the production term, the closure of the Reynolds tensor in first gradients is *saturated*, because it is limited by the realizability conditions. Moreover, there is no term explicitly dependent on the artificial viscosity in the equations of $\tilde{k}$ and $\tilde{\varepsilon}$ (DAM's model) or $\tilde{k}_b$ and $\lambda_i = (C_i/C_\mu)\lambda_d$ (AWE's model). The equation system of the turbulent quantities in the *Lagrangian* (non-Galilean) coordinate system of the fluid thus becomes:

$$\frac{\mathrm{d}}{\mathrm{d}t}\tilde{k} = \pi, \tag{8.8a}$$

$$\frac{\mathrm{d}}{\mathrm{d}t}\tilde{\varepsilon} = C_{\varepsilon1}\frac{\tilde{\varepsilon}}{\tilde{k}}\pi - C_{\varepsilon3}\,\tilde{\varepsilon}\,(U_x)_{,x} \qquad \text{DAM's model}, \tag{8.8b}$$

$$\frac{\mathrm{d}}{\mathrm{d}t}\lambda_i = \left(-\frac{1}{3} + C_\lambda\frac{2\cos^2\psi - \sin^2\psi}{3}\right)\lambda_i\,(U_x)_{,x} \qquad \text{AWE's model}, \tag{8.8c}$$

where, for AWE's model, $\tilde{k}$ is identical to $\tilde{k}_b$, and the equation of λ_i, obtained from the equation of λ_d in (3.32), involves the angle ψ between the directions of the shock and of the volume fraction gradient of the passive fluids in the environment. π is the production term by the modeled Reynolds stresses:

$$\pi = -\frac{\overline{\varrho u_i'' u_j''}}{\overline{\varrho}}\,(U_i)_{,j} \overset{\mathrm{m}}{=} b_{\max}\,\tilde{k}\,\frac{\mathrm{d}}{\overline{\varrho}\,\mathrm{d}t}\overline{\varrho}\,, \tag{8.9}$$

where $b_{\max}$ is the maximum anisotropy ratio, $b = \overline{r_{xx}}/(\overline{\varrho}\tilde{k})$, allowed by the limitation of the closure ($2/3 < b < 2$ in general, and $b_{\max} = 5/4$ in DAM's standard k-ε model [5, 6]). To obtain jump relations, we must show the invariant quantities of equations (8.8a) and (8.8b), which are logarithmic:

$$\frac{\mathrm{d}}{\tilde{k}\,\mathrm{d}t}\tilde{k} = b_{\max}\,\frac{\mathrm{d}}{\overline{\varrho}\,\mathrm{d}t}\overline{\varrho}, \tag{8.10a}$$

$$\frac{\mathrm{d}}{\tilde{\varepsilon}\,\mathrm{d}t}\tilde{\varepsilon} = (C_{\varepsilon1}\,b_{\max} + C_{\varepsilon3})\,\frac{\mathrm{d}}{\overline{\varrho}\,\mathrm{d}t}\overline{\varrho} \qquad \text{DAM's model}, \tag{8.10b}$$

$$\frac{\mathrm{d}}{\lambda_i\,\mathrm{d}t}\lambda_i = \left(-\frac{1}{3} + C_\lambda\frac{3\cos 2\psi + 1}{6}\right)\frac{\mathrm{d}}{\overline{\varrho}\,\mathrm{d}t}\overline{\varrho} \qquad \text{AWE's model}, \tag{8.10c}$$

from which we obtain the turbulent kinetic energy and the integral length scales:

$$\boxed{\begin{aligned} \tilde{k}_1 &= \tilde{k}_0\left(\frac{\varrho_1}{\varrho_0}\right)^{b_{\max}}, \\[4pt] \lambda_{i1} &= \lambda_{i0}\left(\frac{\varrho_1}{\varrho_0}\right)^{(3/2 - C_{\varepsilon1})\,b_{\max} - C_{\varepsilon3}} \qquad \text{DAM's model}, \\[4pt] \lambda_{i1} &= \lambda_{i0}\left(\frac{\varrho_1}{\varrho_0}\right)^{-1/3 + C_\lambda(3\cos 2\psi + 1)/6} \qquad \text{AWE's model}. \end{aligned}} \tag{8.11}$$

For a strong shock in an ideal gas at $\gamma = {}^5/_3$, the compression ratio is close to $\varrho_1/\varrho_0 = 4$. We then observe:

- an increase in the turbulence by a factor of at most $\tilde{k}_1/\tilde{k}_0 = 4^{5/4} \approx 5.66$ in all the models,
- in DAM's k–ε model, the quasi-invariance of the integral length scale using the values of $C_{\varepsilon 1}$ and $C_{\varepsilon 3}$ in Table 2.1, p. 14,[4]
- in AWE's two-fluid model, a variation of the integral length scale between $\lambda_{i1}/\lambda_{i0} = 4^{-(1+C_\lambda)/3} \approx .4$ (transverse gradient of $\alpha^\pm$) and $\lambda_{i1}/\lambda_{i0} = 4^{-(1-2C_\lambda)/3} \approx 1.6$ (longitudinal gradient of $\alpha^\pm$), using the value of C_λ in Table 3.1, p. 27.

The amplifications in (8.11) have indeed been observed qualitatively for DAM's k–ε model in a previous numerical study,[5] where $\varrho_1/\varrho_0 \approx 3.67$ with $C_{\varepsilon 3} = .4$ and $b_{\max} = 1.6$ [26, §4.1]. For AWE's two-fluid model, we must point out that it behaves very much like the k–ε (the equation in $\tilde{k}$ is identical and that in $\tilde{\varepsilon}$ is replaced by a practically equivalent equation in λ_i), but is inconsistent, because its response depends on the orientation of the gradient of a passive quantity.

These results would be acceptable in the application range of turbulence models, that is, for a distortion that is slower than or of the order of the integral time scale $\Delta t > \lambda_i/\tilde{k}$. Actually, the characteristic shock-compression time is linked to the dissipative mechanism (for example, the viscosity) and is very much smaller than the shortest turbulence scale, the Kolmogorov time. With these conditions, the turbulent field can be considered as frozen when the shock passes, and the shock–turbulence interaction is well described theoretically by a linear analytic development [150], known as Linear Interaction Analysis (LIA).[6] The LIA results have been successfully verified by direct simulations [152–157], and thus *we shall adopt the LIA as a reference for the modeling*. As a function of the Mach number, this theory leads to amplification factors of which we shall retain the following properties in the far field:

- $\tilde{k}_1/\tilde{k}_0$ is greater than 1, increasing, and asymptotically approaching about 1.75 (practically reaching it at Mach 3);
- beginning at about Mach 1.5, λ_1/λ_0 is less than 1, decreases, and asymptotically approaches 0.

Thus we see a significant divergence of the k–ε type models from the LIA reference. To restore the asymptotic behavior of $\tilde{k}_1/\tilde{k}_0$, one would have to

---

[4] A more classical value of $C_{\varepsilon 3} = {}^1/_3$ would lead to a decrease of λ_{i1} down to the limit $\lambda_{i1}/\lambda_{i0} \approx .63$.

[5] Most of the numerical schemes, which are conservative in k and ε, cannot *precisely* reproduce the jump relations (8.11). Considering the nature of the evolution equations (8.10a) through (8.10c), they would need to be conservative in $\log \tilde{k}$ and $\log \tilde{\varepsilon}$ in order to do that.

[6] We will not consider rapid distortion theory (RDT) here [151], because it does not apply to unconfined shocks.

select, for example, $b_{max} \approx .4$, which is incompatible with the constraint $b_{max} > 2/3$ stemming from its definition. Regarding λ_1/λ_0, the asymptote cannot be reconstructed, regardless of the values of $C_{\varepsilon 3}$ or C_λ.

The major difference between the behaviors predicted by LIA and by equations (8.11) thus shows that k–ε type modelings *must be specifically corrected for the presence of shocks*, as for example in the VIKHR model [13–17], using artificial dissipative terms. An approach well adapted to DAM production codes would be to add terms related to the shock force, characterized by the quantity $Q/(\varrho c_s^2)$, where Q is the artificial viscosity.

9

Some Perspectives on Further Developments of Models[1]

9.1 About Second-Order Extensions of Single-Fluid Turbulent Mixing Models

Faced with the insufficiencies of the basic single-fluid models, such as DAM's k–ε, several authors [13–26,167] have rightly identified the source of the problems as being in the modeling of the turbulent mass flux, $\overline{u}''$, which appears in the so-called "Rayleigh–Taylor" source term. Their strategy then has been to better capture this term by introducing a single-fluid equation for the evolution of $\overline{u}''$. Since this quantity corresponds to a second-order correlation, $\overline{\varrho' u'}$, modeling consistency also requires that evolution equations be added for the Reynolds tensor and, especially, the density variance $\overline{\varrho'^2}$. This last quantity is important because it specifically appears in a $\overline{\varrho u_i''}$ production term having the form $\overline{\varrho'^2}(\overline{p})_{,i}/\overline{\varrho}^2$. The VIKHR [13–17], BHR [22–25], and MELT-2O [26–28] models are based on this approach, although they differ in many respects.[2]

---

[1] © This chapter has significantly changed compared to the original version. The original second section, "About drag closure in AWE's two-fluid model," is now irrelevant in view of recent advances in the phenomenology of growth and friction of buoyant structures [164–166], and has thus been completely removed. A new section has been added to take into account some important results on SSVARTs obtained in the aftermath of the original work [54,55], which significantly affect the conclusions and future modeling strategies.

[2] In the VIKHR [13–17] or BHR [22–25] models, the quantities $\overline{u}''$ and $\overline{\varrho'^2}$ are denoted, respectively, as $\boldsymbol{W}$ or $\boldsymbol{A}$ and $\overline{\varrho}B$ or $\overline{\varrho}^2 B$.

Moreover, in the VIKHR model, the Reynolds tensor is defined starting from velocity fluctuations *relative to the Reynolds average*, not the Favre average; there are thus corrective terms in $\boldsymbol{u}''$, but in the limit of zero Atwood number, a calculation similar to that in Sect. 7.3 shows that R_{ij} has no contribution from the directed kinetic energy.

Finally, in addition to global consistency to second-order, the introduction of R_{ij} is also justified by the need to reconstruct the relaxation of the Reynolds tensor asymmetry created by the interaction with a shock [167]. Considering the arguments

A. Llor: *Statistical Hydrodynamic Models for Developed Mixing Instability Flows*,
Lect. Notes Phys. **681**, 107–117 (2005)
www.springerlink.com   © Springer-Verlag Berlin Heidelberg 2005

We shall not make a "0D" analysis of these models here, which would be time-consuming, although not inherently difficult. We prefer instead to establish a link with the two-fluid approach in order to estimate the potential benefit of these second-order models. The Reynolds tensor and the turbulent mass flux have already been expressed in two-fluid terms in (7.6). In the same spirit of separating directed and random parts, the variance of density fluctuations in the mixture decomposes as:

$$\overline{\varrho'^2} = \overline{(c^+ + c^-)(\varrho - \overline{\varrho})^2} = \overline{c^+(\varrho - \varrho^+ + \varrho^+ - \overline{\varrho})^2} + \overline{c^-(\varrho - \varrho^- + \varrho^- - \overline{\varrho})^2}$$
$$= \overline{c^+}(\varrho^+ - \overline{\varrho})^2 + \overline{c^-}(\varrho^- - \overline{\varrho})^2 + \overline{c^+(\varrho - \varrho^+)^2} + \overline{c^-(\varrho - \varrho^-)^2}$$

$$\boxed{\overline{\varrho'^2} = C^+ C^-(\varrho^+ - \varrho^-)^2 + \overline{c^-(\varrho - \varrho^-)^2} + \overline{c^+(\varrho - \varrho^+)^2}.}$$
$$(9.1)$$

As was the case for the expressions of turbulent fluxes in (7.6), $\overline{\varrho'^2}$ contains two contributions, due to:

• the difference between the mean densities of the fluids, and
• the density fluctuations within each fluid relative to their mean densities.

Thus we see that the second-order single-fluid approaches involve, implicitly through $\overline{u}''$ and $\overline{\varrho'^2}$, the velocity and density differences between the fluids, and thus reduce to a two-fluid approach. It is thus reasonable to expect these models to have the same capabilities of capturing the instability parameters examined in Chaps. 4 through 7.

Despite this potential, the second-order single-fluid approaches only partially address the various difficulties of the first-order single-fluid models discussed in Chap. 7. To summarize:

1. **The equations in R_{ij}** do not appear to be mandatory, because an algebraic closure, possibly derived from an ASM-type analysis [146], would make it possible to capture the anisotropy of the Reynolds tensor [33, 128–130], even within the framework of a first-order single-fluid model. Thus it would be possible to simplify many tricky aspects of these equations: separation of the directed and random contributions, adaptation to the treatment of shocks, realizability, numerical stability, etc.

2. **The directed and turbulent components** are not separated in the definitions and closures of the various turbulent quantities. Thus difficulties can be expected when the ratio K/K_D is small or variable, or else when K and K_D are associated with very different characteristic scales (RT instability with variable acceleration, demixing, etc.).

3. **Modeling of the "Rayleigh–Taylor" term of k production** does not necessarily appear to be correct in the available second-order models. This modeling is based on the *same* statistical analysis as for the first-order

---

presented in Chap. 8, this effect does not appear dominant when describing mixing zone growth, and it can be taken into account, at least partially, by an adapted model of algebraic closure and shock treatment.

models, which were shown in Sect. 7.2 to be inadequate for flows where the density variations are due to the effects of both mixing of distinct fluids and compressibility. As was shown in Sect. 7.2, this can cause a violation of the second law of thermodynamics, such as observed in certain cases with DAM's k–ε model. The reconstruction of self-similar Rayleigh–Taylor instabilities with acceleration varying as t^n, as introduced in Sect. 5.3, thus constitutes an important validation test for these models.

The multiscale extension of turbulence models has also been proposed as a possible means of improvement [26,167], but this also encounters the obstacles described in items 2 and 3 above.

9.2 Molecular Interdiffusion, Demixing, and Separation of Geometric and Turbulent Scales

9.2.1 Interdiffusion[3]

In a mixing zone, turbulent stirring causes fragmentation of the fluids at all scales and thus produces a distribution of structure sizes that is not described by the drag scale λ_d alone. At smaller scales and for miscible fluids, molecular interdiffusion even leads to an intimate mixture. Because the drag between the fluids varies inversely as the size of the structures, the small inclusions of one fluid in the larger structures of the other are thus entrained by the latter. The dynamics of the mixing zone growth is thus controlled by the size of the largest structures: since these are not pure, their densities are not those of the original fluids, and consequently the effective values of the Atwood number and drag must be modified. This small-scale mixing of fluids by turbulence will be called *interdiffusion*, which is understood to be turbulent, even if, strictly speaking, there is no molecular interdiffusion in the case of immiscible fluids.[4]

AWE's two-fluid model considers that the fluids are immiscible and preserve their initial properties, particularly their densities, at all times, but this does not seem to produce inconsistent results. Actually, because of the nature of the investigated flows, which are self-similar and in the limit of zero Atwood number, the effects of interdiffusion can be absorbed by the various constants of the model. However, for more general flows, especially in demixing phases, interdiffusion must be taken into account [51,169,170].

The influence of interdiffusion has been clearly discussed on a phenomenological basis by G. Dimonte [51], who emphasized the importance of faithful reconstructions *as a function of the Atwood number*. In this approach, mentioned in Chap. 1 (see footnote 3, p. 3), it is thus mandatory to discuss interdiffusion effects. Let us review the conclusions of this notable work:

---

[3] (T) The last paragraph of this part was added to the original version.

[4] (T) Interdiffusion should be more appropriately designated by "mass exchange" [168].

1. The author's aim was to use a simple phenomenological model to reconstruct the mixing zones observed over a broad set of experiments involving RT instabilities obtained using an accelerated projectile in the Linear Electric Motor (LEM): constant-acceleration RT, incompressible RM, and RT with variable (but not self-similar) accelerations, in each case with Atwood numbers ranging from .1 to .9.

2. All the models considered could be reduced to very simplified "double-0D" momentum equations. The term "double-0D" refers to averaging all quantities separately over the two halves of the mixing zone (divided by the initial position of the interface) in order to capture the growth behavior of the two edges of the TMZ with coupled ODEs. This simple "double-0D" system reduces to:[5]

$$\frac{\mathrm{d}\dot{L}^\pm}{\mathrm{d}t} = C_a \mathcal{A}\, \Gamma(t) - C_d\, \frac{|\dot{L}^\pm|\,\dot{L}^\pm}{L^\pm}\,, \tag{9.2}$$

where the $L^\pm$ mark the mixing zone edges relative to the initial position of the interface. This makes it possible to capture very asymmetric situations, when the Atwood number is close to 1.[6]

3. Various published phenomenological models derive from this general framework [49, 50], but with different algebraic closures of C_a and C_d (buoyancy and drag coefficients). These closures, which were obtained heuristically, involve the densities of the fluids and attempt to reconstruct the behaviors as a function of the Atwood number and the acceleration profile. The two limits of Atwood number equals zero (symmetry) and unity (free fall of a heavy fluid) are then particularly constraining.

4. By systematically solving the "double-0D" equations of various models, Dimonte observed large discrepancies compared with the experimental results. In all cases, he was able to attribute these discrepancies to the poor evaluation of the effective densities in the buoyant force and drag terms, which define C_a and C_d.

5. Dimonte thus proposed a closure consistent with the effective densities that define C_a and C_d by taking into account on the one hand the mean density profile, in two linear parts or "double-0D," and on the other hand

---

[5] Ⓥ This is the generic form of the *buoyancy–drag models*. Buoyancy–drag balance in Rayleigh–Taylor flows has been identified by many authors over at least twenty years [51] to be the dominant control mechanism of growth. Many simplified or bulk models have thus been built around this concept, but its importance has been surprisingly underestimated or ignored in the development of "sophisticated" turbulent statistical models, especially extensions of the single-fluid k–ε model. The present report makes no exception and does not take advantage of this powerful concept which is first mentioned here, close to the end.

[6] Ⓥ Additionally C_a and C_d are functions, *common* to both + and − equations, of only two variables: the density of the *fluid at rest* on the side under consideration (+ or −), and the *mean density at the center of the TMZ* (which is affected by interdiffusion).

the partial interdiffusion of the fluids. Then by adjusting only one residual parameter characteristic of the drag scale, he was able to obtain a very good agreement with all his experimental results over the entire range of Atwood numbers.

Although turbulence is only implicitly taken into account, through the coefficient values, it is noticeable that an excellent agreement is obtained for the influence of the Atwood number.

This study provided an a posteriori justification for a major advance of AWE's two-fluid model previously introduced by D.L. Youngs [168]: the addition of a mass exchange term between the fluids. This term, constructed in an entirely phenomenological way, was adjusted by a dimensionless coefficient to reconstruct the evolution equation for the variance of a passive scalar in the limit of zero Atwood number. Rather surprisingly, neither the value of the coefficient nor even the complete equations of the model seem to have been published. This model appears to correctly reconstruct variable-acceleration experiments [168] and particularly demixing (where L decreases asymptotically toward a non-zero value).

Taking into account interdiffusion (or mass transfer) has two major consequences regarding the "0D" analysis of the energy balance given in this report:

- The mutual entrainment of the fluids makes their mean velocities (as given in (4.3)) to be averages over fluids which are mixed into light upward-moving and heavy downward-moving *entities*. The entrainment of the fluids, as characterized by the so-called "molecular mixing" coefficient [168–170], is strong enough to make the effective drift velocity between *entities* about 15% higher than the drift velocity between *fluids* δU. With this new approach, the corresponding directed energy makes up about one third of the total turbulent energy (in contrast to only one fourth for the two-fluid analysis in Sects. 4.2 and 4.3). The discussion in Sect. 7.3 on the orders of magnitude of the two-fluid contributions to the single-fluid turbulent fluxes should be corrected accordingly.

- More concerning is the fact that the internal dynamics of the entities inside the TMZ cannot be captured by just giving the TMZ growth rate and the volume fraction profiles of the fluids as in Sect. 4.2. Supplementary information is required to characterize the fluid entrainments, for instance with effective density profiles. The corresponding "0D" analysis may then be significantly more complex than what is shown here [171]. Nevertheless, the conclusions of the report do hold even if they may appear as insufficiently restrictive for modeling.

9.2.2 Geometric and Turbulent Scales[7]

In an RT instability at constant acceleration, only one characteristic length scale, $\mathcal{Y}A\Gamma t^2$ exists. The ratios between the mixing zone thickness, the integral length scale of turbulence, and the characteristic drag length, L, Λ_i, and Λ_d, are thus well defined and constant. In usual textbooks, it is commonly assumed that the Λ_i/L ratio is approximately identical in all mixing layers, because of the causal relationship between turbulence and TMZ growth. However, when the acceleration varies, this assumption is clearly incompatible with the elementary data gathered here: there is a significant difference in von Kármán numbers between the RT and RM instabilities. The relationship between TMZ growth and turbulence actually involves only a part of the latter, namely the directed kinetic energy K_D, and the ratio K/K_D can vary over orders of magnitude.

Similarly, the Λ_d/L ratio has no reason to be fixed. Yet in Dimonte's "double-0D" phenomenological analysis, the Λ_d/L ratio is fixed (despite the author's awareness of the difficulty). More surprisingly, this ratio is also fixed for AWE's two-fluid model according to (6.2), notwithstanding the presence of a λ_d equation in the model which, in principle, could have captured possible changes.

Just as striking is the fixed value of the Λ_d/Λ_i ratio in AWE's two-fluid model, as given by the algebraic closure in (3.33d). Yet one can imagine extreme situations of opposite behaviors of the turbulent and geometric scales: for example, in the case of demixing by inversion of the acceleration after a period of instability, there could be a slight decrease in the mixing zone thickness, limited by the integral length scale which can only increase, together with a strong reduction of the drag scale, reflecting the break up of fluid structures under the effect of turbulent stirring.

Therefore, in order to capture the effects of randomly variable accelerations, it seems as important to enrich AWE's two-fluid model by dropping the direct relationship between L, Λ_i, and Λ_d, as to introduce mass exchange between fluids as mentioned above [168]. Naturally, this requires a serious reassessment of the λ_d equation and the addition of an evolution equation for the turbulent length scale, for instance through an ε equation.

A new model development was thus attempted at the DAM along these lines. Starting from the usual two-fluid equations, a k–ε model was added, and for drag closure, λ_d was substituted by Σ (the interface area per unit volume) for which an evolution equation was provided according to available closures of statistical equations [88]. This approach, called k–ε–Σ [52,53], failed but provided important insight for developing a consistent approach. The basic flaw in the k–ε–Σ model was the inconsistent profiles of the length scales λ_i

---

[7] ⓊThis part has been substantially expanded in order to take into account the main findings of our recent works [52,53,164–166]. The original version had been kept deliberately short and somewhat sibylline in order to preserve the, at the time, new approaches and ideas that were being investigated.

and λ_d across the TMZ: as with AWE's two-fluid model, λ_d had to be roughly uniform, whereas, as with any k–ε model, λ_i had to be bell shaped and vanish at the TMZ edges. Through couplings between these two length scales in a variety of terms, the model equations always produced unphysical singularities at the TMZ edges.

A new approach was then proposed, denoted 2SFK [164–166], where the two length scales λ_i and λ_d are identical and merged into a single modeled parameter, as supported by experimental evidence. The evolution of this common length scale is then controlled by two sets of coupled k–ε equations in order to allow non-vanishing integral length scales at the TMZ edges. Interestingly, the buoyancy-drag balance and mass transfer process are easily included into this modeling concept. Although the first versions of our 2SFK models were far from optimal, the von Kármán numbers of the Rayleigh–Taylor and Richtmyer–Meshkov flows were readily captured without *any* adjustment of model coefficients [164–166]. This significant improvement compared to previous models supports the general 2SFK concept even if detailed closures can be improved. Also noticeable is that the model was easily adapted to capture the enhanced diffusion process that has recently been shown to exist in RT layers [171].

9.3 ⓤ Self-Similar Variable Acceleration Rayleigh–Taylor Flows (SSVARTs)[8]

9.3.1 Extension of SSVARTs to $n < -2$

Self-Similar Variable Acceleration Rayleigh–Taylor flows (SSVARTs), for which acceleration varies as $\Gamma(t) \propto t^n$ as considered in Sects. 5.3, 6.2, and 7.4, have actually been introduced in earlier publications, but without a full assessment of their potential for model testing and with the restriction $n > 0$ or $n > -2$ [48,172]. In fact, they can be extended below $n = -1$, the limit considered here so far, thus providing new insights and more stringent modeling constraints [54,55].

In order to generate the most general SSVARTs, one must consider what are the minimal basic constraints to be fulfilled by such flows. We shall retain the following two conditions which are closely related to the mixing irreversibility (or the second law of thermodynamics):

- $L(t)$ is positive and always grows, because demixing is unphysical (it would proceed towards a *vanishing* TMZ width to ensure self-similarity, thus separating molecularly mixed fluids), and

---

[8] ⓤ Shortly after the original version of this worked appeared, it was realized that SSVARTs could be extended to $n < -2$ [54,55]. This development and its consequences are so closely connected to the present work that they have been added in this new section.

- $L(t)$ must be asymptotically stable with respect to small changes of initial conditions, in particular of the initial directed energy.

From $\Gamma(t) \propto t^n$ for $t > 0$, one infers $L(t) = \mathcal{Y}_n \mathcal{A} \Gamma(t) t^2 \propto t^{n+2}$ with $\mathcal{Y}_n > 0$, and the growth condition yields $n > -2$. Now, growth and self-similarity are also ensured if:

$$\Gamma(t) \propto (-t)^n \text{ for } t < 0 \text{ with } n < -2 , \tag{9.3}$$

so, as required, $L(t) = \mathcal{Y}_n \mathcal{A} \Gamma(t) t^2$ is positive and grows. In this case, the "initial" time, corresponding to a vanishingly perturbed flat interface, is rejected to $t = -\infty$ and the TMZ width diverges in a finite time when approaching $t = 0$.

Regarding the initial condition constraint, one can consider it is fulfilled whenever the input of potential energy is sufficient to eventually supersede the (diluted and partly dissipated) initial energy perturbation. For $n < -2$ the self-similar energies always grow and overrun the initial perturbations. For $n > -2$ the self-similar energies in a SSVART vary as $K \propto t^{2(n+1)}$, whereas the initial energy perturbation is diluted and dissipated according to $t^{-p_n}/L \propto t^{-n-2-p_n}$, where p_n is a positive number to take dissipation into account. Therefore, initial conditions are "forgotten" if $n > -4/3 - p_n/3$.

The RM flow actually provides a reference to define the threshold $n = -4/3 - p_n/3$. For the RM case $L(t) \propto t^{n_0}$ where $n_0 \approx .3$, and thus whenever $n < -2 + n_0$ the initial energy in the mixing zone will eventually dominate the input energy. SSVARTs are thus limited to $n > -2 + n_0$ and $n < -2$, the $0 < n < -2 + n_0$ gap representing an unstable zone or an attraction basin towards the RM flow [54, 55].

After lengthy and tedious though straightforward calculations, the responses to SSVARTs at $n < -2$ of all the models considered in this study turn out to be formally identical to those at $n > 0$ given in (5.11), (6.12), and (7.12) [54, 55]. Remarkably, according to these formulas, every model yields a singularity at $n = -2 + n_0$ where n_0 is the growth exponent of the RM instability *as predicted by the same model*. There is thus self-consistency of all the models with respect to this boundary of the forbidden range of n values (related to the energy constraints in the RM limit). All models, except DAM's k–ε however, also yield à singularity at $n = -2$, in accordance with the n gap limits found above.

Most interestingly for practical purposes, the expansion in the range of n values further constrains the models: in Sect. 5.3 the stability domain for the $C_{\varepsilon 0}$ and σ_ϱ coefficients of DAM's k–ε model is more restricted, leading in particular to $C_{\varepsilon 0} > 3/2$ instead of $C_{\varepsilon 0} > 1$. However, for $C_{\varepsilon 0} = 3/2$ and $\sigma_\varrho = .7$, the growth rate of the RT TMZ is about 3% of the experimentally observed value. Moreover, in contrast to what was stated in Sect. 7.4, drift models (including "Read's formula" in (7.16)) also appear to behave unphysically by overrunning free fall for $n \to -2^-$ and should thus be discarded. Full two-fluid models, such as AWE's, display acceptable responses over the full range of n values (see figure in [54, 55]).

By expanding the range of n, SSVARTs span the three flow regimes that can be considered when each of the three terms in the buoyancy–drag equation is assumed to be negligible [54,55]. Free fall and Read's formula then appear as two extreme cases which bound the behavior of any model [54,55].

9.3.2 Spectral Quasi-Equilibrium and Relevance of Models in SSVARTs

At the moment, until experiments or simulations will be performed, SSVARTs must be considered as "thought experiments." As such, there could be some debate, or even suspicion, on their status as flow references for modeling, and this issue deserves a few comments to further the discussions given in Sect. 5.3 and in [54,55]. For that purpose, five aspects will be distinguished here which have not been explicitly examined so far:

- the existence of a mean flow generated as the response of the physical system to a self-similar forcing in $\Gamma \propto t^n$,
- the nature of the velocity fluctuations and of their dissipation (turbulent energy spectrum and inertial range) around the usual RT case at $n = 0$,
- the possible behavior of the turbulent energy spectrum for large n,
- the significance and relevance of model responses to SSVARTs.
- the robustness of model responses to SSVARTs,

Existence of SSVARTs. Independently of any relevance and description concern, it is *always possible* to consider RT mixing flows driven by $\Gamma \propto t^n$ where $-\infty < n < \infty$. It is also *always possible* to consider the corresponding *Favre ensemble averaged flows*. Assuming *self-similar response* it is also possible to analyze the dependence on energy initial conditions as shown above and determine the gap of forbidden n values. Yet, the *relevance* of the mean flow is not fully ensured: indeed, one could imagine the highly unlikely situation where, at fixed n, the evolution of a *given realization* would be locked on either of two (or more) different TMZ growth coefficients. Within the range of n values which would produce these phenomena of "long intermitency," "symmetry breaking," or "ergodicity loss," the ensemble averaged flows would be of little relevance with respect to individual realizations (and to engineering needs). Such situations could appear for steep growths of Γ if various turbulent structures were stable (see below).

Turbulence in SSVARTs Around $n = 0$. Once the mean flow is known and the system assumed to be ergodic (see above), it is *always possible* to obtain fluctuations around averages and the corresponding power spectra. Self-similarity will also ensure that power spectra evolve self-similarly (always assuming high Reynolds number) and can be collapsed into n-dependent but *constant* shapes.

By further *assuming* that power spectra display self-similar profiles over some range in $\mathbf{k}$ space (wave number) as $E(|\mathbf{k}|) \propto |\mathbf{k}|^p$, bounds on p can be discussed. For instance, the usual quasi-equilibrium between large-scale production and small scale dissipation leads to a Kolmogorov spectrum with $p = -5/3$, but, at the other extreme, if fluctuations produced at large scales remain frozen at these scales at later times as the TMZ grows, then:

$$p = -\frac{3n + 4}{n + 2} , \qquad (9.4)$$

The Kolmogorov spectrum is thus mimicked for $n = -1/2$, and for $-4/3 < n < -1/2$ the frozen modes assumption yields $0 > p > -5/3$: these values of p are clearly forbidden because the characteristic time scales within such a spectrum would very quickly relax it towards the Kolmogorov $p = -5/3$ and production could not overrun dissipation at low n. For $n = 0$ (RT case), the frozen modes assumption yields $p = -2$, barely below the Kolmogorov value (and even closer if taking into account intermittency corrections), and for $n = \infty$, $p = -3$. Estimation of the integral length scale according to (4.10) is therefore perfectly valid even for n values somewhat above 0.[9]

Turbulence in SSVARTs at Large n. The frozen modes assumption considered above provides some insight for steep Γ evolutions, at large n. If some (partial) spectral quasi-equilibrium is to exist over a limited $|k|$ range, its dynamics must be sufficiently fast so as to at least follow the growth of production. Now time scales in the limits of frozen modes and quasi-spectral equilibrium vary respectively as:

$$\tau(|k|) \propto \begin{vmatrix} (\Gamma/t^n)^{-1/(n+2)} \, |k|^{-1/(n+2)} , \\ \varepsilon^{-1/3} \, |k|^{-2/3} . \end{vmatrix} \qquad (9.5)$$

For sufficiently large $|k|$, the usual Kolmogorov cascade could always overrun the frozen modes profile provided that $n > -1/2$. In the spirit of footnote 8, p. 42, steep accelerations could induce strongest departures from the Kolmogorov cascade at small length scales: the sub-inertial range would widen as n increases (intermitency would also increase), thus confining the inertial Kolmogorov cascade to ever smaller scales. This tentative analysis provides a rationale for the small value of the von Kármán number in the RT flow.

Significance of Models for SSVARTs. The analysis given above shows that models requiring full spectral quasi-equilibrium (as the k–ε) may capture accurately the SSVART flows only for small values of n. However, if the sub-inertial range is properly taken into account by a two-fluid description (or maybe a multiscale approach) while a k–ε approach takes care of the inertial range, one can expect the range of acceptable n values to expand significantly above 0. The 2SFK model recently developed at the DAM follows this general scheme [164–166]. Precise n limits on these modeling approaches could be defined from future experiments or numerical simulations.

---

[9] ⓤ Incidentally, the time behavior in the KH case is identical to that of a SSVART at $n = -1$, with dissipation ε (and production) diminishing as $1/t$, to be compared to a growth as t in the RT case and a constant ε in a SSVART at $n = -1/2$.

Robustness of Models to SSVARTs. Even if the accuracy of a given model is out of reach on SSVARTs with large n, its *stability and robustness should be ensured over the full range of acceptable n values*. Indeed, numerical trials on DAM's k–ε model with $C_{\varepsilon 0} = 1$ (as recommended in Sect. 5.3) were found unstable in applications with strong accelerations changes and shocks, although they have been shown here to be acceptable for SSVARTs at $n > 0$. With $C_{\varepsilon 0} = 3/2$, as found when taking also into account SSVARTs at $n < -2$, full stability was then obtained albeit with unphysically small growths of the TMZ. A rationale for the importance of SSVARTs at $n < -2$ is that, in the simplified buoyancy–drag picture, the inertial regime where drag is negligible is obtained for SSVARTs when $n \to -2^-$: this inertial regime is precisely expected during transients with steep acceleration increases. Whatever the behavior of turbulence in this regime and its modeling, the TMZ growth in the inertial regime must be bounded by free fall.

Summary of Part III

This part was devoted to exploring generic ideas for modeling, for instance in the light of the model-specific "0D" responses obtained in Part II. Three main modeling aspects have been reviewed: directed effects in single-fluid approaches, interactions between shocks and heterogeneities, and the related phenomena of fluid entrainement and length scale descriptions.

Poor performance on RT and SSVART flows of single-fluid models compared to two-fluid models is not specific to DAM's k–ε and AWE's two-fluid. Fundamental difficulties of single-fluid approaches can be traced back to the decomposition of the statistical term of "enthalpic" or "Rayleigh-Taylor" production, and to the directed contributions in the turbulent fluxes. Using these analyses it is possible to somewhat correct the algebraic closures of single-fluid models. These approaches can be seen as "extended single-fluid" or "degenerated two-fluid." However they are still found to be *unphysical for some SSVARTs*, leaving only full two-fluid approaches as generally acceptable.

As for the mixing instabilities in the incompressible limit, the behavior of models in the presence of shocks must be assessed as a function of a predefined specification list. A preliminary analysis provides a minimal approach, whereby details of pressure relaxation after shock passage are ignored and overall jump relationships are given for the quantities in the turbulent modeling (without addition of new evolution equations). In the limiting case of the interaction of a shock with a statistically homogeneous turbulent medium, analytical estimates can be obtained for DAM's k–ε and AWE's two-fluid models to be compared with results of an analytical theory (LIA): similar inaccurate behaviors are found for both models. Adapting *artificial dissipation terms* should in principle permit correcting the response of two-fluid models.

Improved approaches are expected from taking into account the mutual entrainment of fluids (through so-called mass transfer terms) and the relationship between the different length scales (TMZ width, integral length scale and drag characteristic length). SSVARTs have also been extended to negative values of n, yielding more straining constraints on models, which will thus need to better capture mass transfer and length scales.

Conclusion, Future Modeling Trends

The relevance of this study stands not in the topics that have been discussed, since most of them have already been explored for many years now, but rather in assembling a broad array of experimental data and hydrodynamic models from the literature under one unified formalism and in subjecting these to a list of evaluation criteria corresponding to the specific needs of the DAM. In this spirit, this study made it possible to define effective approaches for developing and validating new hydrodynamic models for mixing flows.[1]

The four most important evaluation tools developed in this study are:

1. **the "0D" analysis of mixing layers** of the three instabilities, Kelvin–Helmholtz, Rayleigh–Taylor, and Richtmyer–Meshkov, in the self-similar, incompressible limit at zero Atwood number,[2] which makes it possible to estimate the performance of models at very low cost, of the order of a few engineer-weeks, without the prerequisite insertion into a numerical code, compared with several man-years previously (see Chap. 4, and Sects. 5.2, and 6.1);

2. **the turbulent structure and equilibrium between the different types of energies** in the mixing layers, characterized particularly by the *von Kármán number* and the proportion of *directed kinetic energy*, which the models must correctly reconstruct (see summarized results in Tables 4.1, p. 38, 5.1, p. 52, and 6.1, p. 68);

3. **the Rayleigh–Taylor instability with self-similar acceleration** varying as t^n, the consideration of which makes it possible to estimate the performance of models in situations close to real applications, less

---

[1] ⓒ It must be noticed that giving an explicit "specification list" for validating models as in Chap. 1 is somewhat uncommon, and consensus upon its detailed content is not yet established in the RT turbulence modeling community. The present study should contribute to bringing this crucial issue into the limelight.

[2] The extension to any Atwood number is tedious, but does not involve any inherent technical difficulty.

A. Llor: *Statistical Hydrodynamic Models for Developed Mixing Instability Flows*, Lect. Notes Phys. **681**, 121 123 (2005)
www.springerlink.com

"smooth" than academic flows (see for example the case of the k–ε model in Fig. 5.2, p. 57);

4. **the "0D" analysis of the interaction between a plane steady shock** and velocity or density heterogeneities in a statistically homogeneous medium (see Chap. 8), which makes it possible to validate the formulation of the artificial dissipation terms contained in the models and codes by comparison with the analytic LIA theory [150].

The relevance of these tools was shown using two models which represent common approaches proposed in the literature: DAM's single-fluid k–ε model and AWE's two-fluid model.[3]

The "0D" analysis shows that the k–ε model with a single set of coefficients cannot reconstruct the growth rate observed in the mixing zone of a Rayleigh–Taylor instability while remaining stable in the presence of accelerations varying as t^n (see Fig. 5.2, p. 57 and the boxed information in p. 60-61). This shortcoming is due to the structure of the energy dissipation circuit in the model, which permits the violation of the second law of thermodynamics for certain combinations of coefficients and the acceleration exponent, n. The standard set of coefficients currently recommended for the k–ε model thus leads to instabilities when $n > .89$. Values can be assigned that ensure stability for all n, but these lead to a growth rate of the Rayleigh–Taylor instability that is too low by about a factor of three. Conversely, AWE's two-fluid model is satisfactory in all the situations, although the integral length scale, which represents the characteristic size of the turbulent eddies, is somewhat distorted, the von Kármán number being too high by about a factor of four (see Sect. 6.2). More generally, the single-fluid approaches pose problems in reconstructing Rayleigh–Taylor instabilities because they are incapable of correctly capturing the evolution of the directed kinetic energy (see Table 7.1, p. 80). By contrast, with regard to shock–heterogeneity interaction, the two models have comparable deficiencies, producing inconsistent jump relations as illustrated here in the particular case of shock–turbulence interaction (see (8.11)).

This analysis revealed possible paths for the development of new modelings:

1. **the degenerate two-fluid "drift"** or single-fluid extended approach, for example, as explored by VNIITF [18–21], which has many of the

---

[3] Ⓤ Keeping in mind pedagogical goals, specific terms were coined to designate the three main quantities for model evaluation in order to stress their importance: directed energy, "0D" averaging, and von Kármán number. Of these only the first two seem to have gained some acceptance with other investigators in the field. Added in later publications is the acronym SSVART, for Self-Similar Variable Acceleration Rayleigh-Taylor flow.

advantages of the two-fluid models at a very low adaptation cost to existing single-fluid codes (see Sect. 7.4);[4]

2. **the separation of turbulence and dispersion scales** in the two-fluid approaches, for example by introducing an equation in ε into AWE's model, which allows a better capturing of the mixing zone evolution in the presence of variable accelerations, demixing episodes, or combinations of instabilities (see Sect. 9.2);[5]

3. **the ad hoc construction of artificial dissipative terms**, which allows precise capturing of jump relations of various quantities in a two-fluid medium subjected to shocks (see (8.6)).

By contrast, the pursuit of second-order or multiscale modeling should not be a priority. The validation of improved models will require a complete set of calibration data (from experimental, analytic, or numerical sources) regarding two important aspects:

- the Rayleigh–Taylor instability with self-similar acceleration varying as t^n, and
- the interaction between a shock and a two-fluid medium in the limit of weak heterogeneities.

All these aspects are currently being studied in collaboration with various outside organizations, such as the IMFT in Toulouse, the ENSC in Cachan, and the AWE in Aldermaston [57]. Specific experiments are also foreseeable at DAM (shock tube) and with VNIITF in Snezhinsk (gas-gun accelerated vessels).

---

[4] ⓤ This suggestion is now irrelevant because, as summarized Sect. 9.3, drift models are unphysical when dealing with SSVARTs where acceleration varies as t^n with $n < -2$ [54, 55].

[5] ⓤ This suggestion has notably evolved, as summarized in the updates to Sect. 9.2. In fact turbulence and dispersion scales must stay identical, but they have to be separated from the width of the TMZ. A new model was recently developed along this idea [164–166] which successfully captures the basic parameters of an RT mixing layer: growth, molecular mixing ratio and von Kármán number.

A

Notations and Acronyms

With few exceptions, the equations in this report use Einstein's notation for space coordinates and derivatives with respect thereto, with no distinction between covariant and contravariant indices, and with summation on repeated indices. The conservation equations are given in Eulerian (rather than Lagrangian) coordinates in such a way that an ensemble mean can be applied to them. They are written so that they always show the temporal variation and advection to the left of the equal sign, and the sources, fluxes, and exchanges to the right. Finally, we introduced the symbol $\stackrel{\mathrm{m}}{=}$ to indicate physical modelings. This allows making the difference with the strict equality, $=$, and the numerical approximation, $\approx$. Listed below are the symbols used in more than one subsection, grouped according to chapters where they first appear.

Chap. 2: The following symbols represent functions of r (position) and t (time):

ϱ : density,

v : per mass volume,

p : pressure,

e : internal energy,

u : velocity vector, with components u_i,

f : total energy, $f = e + u^2/2$,

τ_{ij} : other-than-pressure stress tensor (opposite sign from conventional); if pure viscous: $\tau_{ij} = -\mu\left((u_i)_{,j} + (u_j)_{,i}\right) - (\lambda - 2/3\mu)\,(u_k)_{,k}\,\delta_{ij}$,

μ, λ : shear and volume viscosities

g : acceleration field

θ : thermal conduction flux,

s : source term of per volume internal energy

$c^{\pm}$: local mass fractions of constituent fluids, (passive values, $c^+ + c^- = 1$),

$\phi^{\pm}$: diffusive mass fluxes of constituent fluids, ($\phi^+ + \phi^- = 0$),

$\bar{a}$: Reynolds ensemble average of quantity a

a' : ensemble fluctuation of quantity a, $a' = a - \bar{a}$,

$\tilde{a}$: Favre average of quantity a, $\tilde{a} = \overline{\varrho a}/\bar{\varrho}$,

A. Llor: *Statistical Hydrodynamic Models for Developed Mixing Instability Flows*, Lect. Notes Phys. **681**, 125–127 (2005)
www.springerlink.com

a" : Favre fluctuation of quantity a, $a" = a - \tilde{a}$,

U : Favre average of velocity $\boldsymbol{u}$,

V : Reynolds average of velocity $\boldsymbol{u}$,

k : *non-averaged* turbulent per mass kinetic energy, $k = u"^2/2$,

ε : *non-averaged* per mass dissipation rate of k, $\varepsilon = -\tau_{ij}(u_i")_{,j}/\varrho$,

r_{ij} : *non-averaged* Reynolds tensor, $r_{ij} = \varrho u_i" u_j"$,

c_s : mean sound velocity,

$C_\times, \sigma_\times$: coefficients of single-fluid model, Table 2.1, p. 14.

We note that only the velocity is defined by a symbol representing the (Favre) average, while all other quantities are used in their non-averaged forms, including k and r_{ij}. This deliberate choice makes it possible to retain the actual significance of each term throughout the calculations.

Chap. 3: Here we add the following symbols and corrections to the single-fluid notations given above:

$\pm$: indices of the two fluids,

$\alpha^\pm$: mean volume fractions of the fluids, $\alpha^\pm = \overline{c^\pm}$,

$C^\pm$: mean mass fractions of the fluids, $C^\pm = \overline{c^\pm \varrho}/\overline{\varrho}$,

$\varrho^\pm$: mean densities of the fluids, $\varrho^\pm = \overline{c^\pm \varrho}/\overline{c^\pm}$,

$\mathcal{A}$: mean local Atwood number, $\mathcal{A} = (\varrho^+ - \varrho^-)/(\varrho^+ + \varrho^-)$,

$\beta^\pm$: distribution coefficients of isentropic compressions, (3.24), p. 24,

$\phi^\pm$: diffusive mass fluxes of the fluids, (thus $\phi^+ + \phi^- = 0$),

$\boldsymbol{U}^\pm$: Favre averaged velocities of each fluid, $\boldsymbol{U}^\pm = \overline{c^\pm \varrho \boldsymbol{u}}/\overline{c^\pm \varrho}$,

$\boldsymbol{u}^\pm$: Favre fluctuations of the velocity in each fluid, $\boldsymbol{u}^\pm = \boldsymbol{u} - \boldsymbol{U}^\pm$,

$\delta \boldsymbol{U}$: drift or interpenetration velocity, $\delta \boldsymbol{U} = \boldsymbol{U}^+ - \boldsymbol{U}^-$,

k_d : mean directed kinetic energy $k_d = C^+ C^- (\delta U)^2/2$,

$k^\pm$: *non-averaged* per mass turbulent kinetic energies in each fluid $k = u^{\pm 2}/2$,

k_b : *non-averaged* two-fluid turbulent kinetic energy, $k_b = c^+ k^+ + c^- k^-$,

$r_{ij}^\pm$: *non-averaged* Reynolds tensors in each fluid, $r_{ij}^\pm = \varrho u_i^\pm u_j^\pm$,

$\varepsilon^\pm$: *non-averaged* volume dissipation rates of $k^\pm$ in each fluid, $\varepsilon^\pm = -\tau_{ij}$ $(u_i")_{,j}$,

$a^\pm$: *non-averaged* Favre fluctuations of quantity a in each fluid, $a^\pm$ $= a - \overline{c^\pm \varrho a}/\overline{c^\pm \varrho}$,

$C_\times$: coefficients of two-fluid model, Table 3.1, p. 27.

We note here that the *lower-case* values indexed with $\pm$ represent Favre fluctuations, except for the eight specific cases of volume fractions, mass fractions, densities, mass fluxes, turbulent kinetic energies, Reynolds tensors, and dissipation rates: $\alpha^\pm$, $c^\pm$, $\varrho^\pm, \phi^\pm$, $k^\pm$, $r_{ij}^\pm$, and $\varepsilon^\pm$.

Chaps. 4 through 6: Here we add the values averaged over the thickness of the mixing layer, as functions of t only and generally shown as upper-case symbols:

L : average thickness of the mixing layer, (4.2), p. 35,

$\mathcal{X}, \mathcal{Y}, n$: self-similar growth coefficients of KH, RT, and RM instabilities,

$\langle . \rangle$: averaging operator over a mixing layer,

K_I : average per mass energy input into a KH layer, (4.8), p. 37, or a RT layer, (4.7), p. 37,

K_M : average single-fluid kinetic energy, $K_M = \langle U^2/2 \rangle$,

K_D : average directed kinetic energy, $K_D = \langle k_d \rangle$,

K : average turbulent kinetic energy, $K = \langle k \rangle$,

K_B : average two-fluid turbulent kinetic energy, $K_B = \langle k_b \rangle$,

E : average per mass dissipation of K into internal energy, $E = \langle \varepsilon \rangle$,

Λ_i : average integral length scale of turbulence, $\Lambda_i = K^{3/2}/E$,

κ : single-fluid von Kármán number, $\kappa = K^{3/2}/(EL)$,

κ_b : two-fluid von Kármán number, $\kappa_b = K_B^{3/2}/(EL)$,

ζ : general correction factor between $\langle a \rangle$ and $a(0)$ (at center), $\zeta \approx 3/2$,

Π_K : average per mass production of turbulence, (5.3), p. 50, and (5.5), p. 51 (DAM's k-ε model),

K_I^* : average per mass energy input into turbulence, (5.6), p. 51 (DAM's k-ε model),

Λ_d : average characteristic drag length, $\Lambda_d = \langle \lambda_d \rangle$ (AWE's two-fluid model),

Π_D : average per mass production of turbulence by drag, (6.6), p. 65, and (6.11), p. 67 (AWE's two-fluid model),

$\mathcal{U}$: ratio of turbulent and directed characteristic velocities, $\mathcal{U} = \sqrt{K_B/K_D}$ (AWE's two-fluid model).

Some acronyms are extensively used throughout this work:

ⓤ : Update (addition or change) to the original version of this report, published in 2001 in French (CEA-R-5983).

ASM : Algebraic Stress Model (closure procedure),

AWE : Atomic Weapons Establishment (UK),

DAM : Direction des Applications Militaires (division of the Commissariat à l'Energie Atomique, France),

KH : Kelvin–Helmholtz,

k-ε : Turbulence model designated by the evolution equations added to the standard hydrodynamic equations,

LIA : Linear Interaction Analysis (shock–turbulence interaction),

RM : Richtmyer Meshkov,

RT : Rayleigh–Taylor,

SSVART : Self-Similar Variable Acceleration Rayleigh–Taylor instability,

TMZ : Turbulent Mixing Zone

0D : Zero Dimensional (i.e. independent of space coordinates).

B

Modeling of Noise Term

An analysis of the so-called "noise" term $\overline{p'(u_i''),_i}$ in (2.11), has been proposed by O. Zeman [148,149] and extended by J.R. Ristorcelli [173], inspired by the decomposition of small perturbations given by L.S.G. Kovasznay [174] and discussed by J. Gaviglio [147]. The approach is to distinguish in the turbulent velocity field, $\boldsymbol{u}''$, the irrotational divergent part $\boldsymbol{u}^d$ and the incompressible rotational part $\boldsymbol{u}^r$, according to the so-called Helmholtz canonical decomposition:

$$\boldsymbol{u}'' = \boldsymbol{u}^d + \boldsymbol{u}^r \quad \text{where:} \quad \begin{cases} \overrightarrow{\mathrm{rot}}\,\boldsymbol{u}^d = \mathbf{0}\ , \\ \mathrm{div}\,\boldsymbol{u}^r = 0\ , \end{cases} \tag{B.1}$$

obtained by the transformation:

$$\boldsymbol{u}^d(\boldsymbol{r}) = -\int \frac{\overrightarrow{\mathrm{grad}}\,\mathrm{div}\,\boldsymbol{u}''(\boldsymbol{s})}{4\pi\|\boldsymbol{r}-\boldsymbol{s}\|}\,\mathrm{d}^3\boldsymbol{s}\ , \tag{B.2a}$$

$$\boldsymbol{u}^r(\boldsymbol{r}) = \boldsymbol{u}''(\boldsymbol{r}) - \boldsymbol{u}^d(\boldsymbol{r})\ . \tag{B.2b}$$

The usefulness of this decomposition stems from the different behaviors of the kinetic energies associated with the rotational and divergent velocity fields [147]:

- in the rotational case, the energy is carried by vorticity, which is locked onto the material in the limit of zero viscosity and weak interactions between eddies [59]. The non-linear interactions between eddies as well as the viscosity introduce a *diffusive* type behavior into the kinetic energy equation;
- in the divergent case, the energy is converted between internal and kinetic energies, which, in the limit of weak perturbations, is equivalent to a process of coherent *propagation*, with characteristic velocities of the order of the sound velocity (acoustic waves). For large compressions, shock waves appear associated with large dissipations, as well as diffractions and dispersions in the presence of heterogeneities or rotational turbulence.

A. Llor: *Statistical Hydrodynamic Models for Developed Mixing Instability Flows*,
Lect. Notes Phys. **681**, 129 130 (2005)
www.springerlink.com

In (2.11), the pressure-related exchange term between internal and turbulent energies $\overline{p'(u_i'')}_{,i}$ is thus associated with the deformations caused by the divergent velocity field *only*, while the dissipative term $-\overline{\tau_{ij}(u_i'')}_{,j}$ contains contributions from *both* fields, if the volume viscosity coefficient, λ, is not zero.

The Helmholtz decomposition of the velocity fluctuations separates two contributions of different types in the turbulent fluxes, in (2.8), and, in particular, reveals *three* components in the turbulent kinetic energy, (2.9b):

$$k = \overline{\boldsymbol{u}''^2}/2 = \overline{(\boldsymbol{u}^r)^2}/2 + \overline{(\boldsymbol{u}^d)^2}/2 + \overline{\boldsymbol{u}^r.\boldsymbol{u}^d} \ . \tag{B.3}$$

In homogeneous isotropic turbulence, the last term averages to zero [148,149], and considering the very different diffusive and propagative behaviors of $\boldsymbol{u}^r$ and $\boldsymbol{u}^d$, we can assume that this property still holds even in anisotropic, heterogeneous, or unsteady turbulence.

However, the absence of correlation between divergent and rotational turbulence is not enough to prove that there is no coupling, that is, no energy *exchanges*, between these two reservoirs. Such exchanges are represented by non-linear terms of order two at least [147], and it seems reasonable to disregard them, *except in shocks* as discussed in Sect. 8.1. Moreover, two situations (found in DAM applications) are appropriate for disregarding divergent turbulence [173]:

- in an open medium, where the high propagation velocity of sound waves rapidly exhausts the reservoir of divergent turbulence,
- when the statistical heterogeneities of the medium are assumed to be in *pressure equilibrium*, which must result in the cancellation of the divergent turbulence field (since it is associated with pressure fluctuations).

Thus in many applications, the term $\overline{p'(u_i'')}_{,i}$ in (2.11) can be neglected. But note that the term $\overline{(p'u_i'')}_{,i}$ must be retained, because it is not necessarily zero for rotational turbulence. It corresponds to the work of the pressure forces due to the effect of turbulence velocities, and thus produces a diffusive type term for the turbulent kinetic energy (a Boussinesq–Reynolds closure can be used) [61, §6.3].

C

Correction of the Input Energy, K_I, of the Kelvin–Helmholtz Instability in the "0D" k–ε Approach

We will now examine the self-similar diffusion equation of a generic quantity a (concentration, $C^{\pm}$, or transverse velocity, U_y) in a Kelvin–Helmholtz mixing layer subjected to turbulent diffusion characterized by a parabolic diffusivity (viscosity) profile over the interval $[-L/2, +L/2]$:

$$\frac{\partial}{\partial t} a = (\nu_t \, a_{,x})_{,x} \; . \tag{C.1}$$

with:

$$a = a(2x/L(t)) = a(\xi) \; , \tag{C.2a}$$

$$\nu_t = \nu_t(0,t) \left[1 - (2x/L)^2\right] = \nu_t(0,t) \left[1 - \xi^2\right] \; , \tag{C.2b}$$

$$\xi = 2x/L(t) \in [-1, +1] \; , \tag{C.2c}$$

$$L(t) = \mathcal{X} \, \Delta U_y t \; , \tag{C.2d}$$

$$\nu_t(0,t) = \zeta \, \frac{C_\mu}{\sigma_a} \frac{K^2}{E} \; . \tag{C.2e}$$

Substituting these expressions into the diffusion equation, elementary calculations then yield an ordinary differential equation in $a(\xi)$:

$$(2\nu_0 - 1)\,\xi\, a' = \nu_0 \left[1 - \xi^2\right] a'' \; , \tag{C.3}$$

where:

$$\nu_0 = \frac{4\nu_t(0,t)\,t}{L^2(t)} = \frac{2\zeta}{\sqrt{3}} \frac{C_\mu}{\sigma_a} \frac{\kappa}{\mathcal{X}} \sqrt{\frac{K}{K_I}} = \frac{\zeta \, \sigma_c}{3\,\sigma_a} \approx \frac{1}{2\sigma_a} \; , \tag{C.4}$$

is a constant coefficient representing the viscosity in the reduced variable ξ, the numerical value of which is given here by taking $\mathcal{X}$, K/K_I, and κ according to Table 5.1, p. 52, and finally $\zeta \approx 3/2$ and $\sigma_c \approx 1$. Equation (C.3) is linear, of first order in a', and has separable variables, so the analytic solution is:

$$a'(\xi) = a'_0 \left[1 - \xi^2\right]^{\frac{1}{2\nu_0} - 1} \; . \tag{C.5}$$

A. Llor: *Statistical Hydrodynamic Models for Developed Mixing Instability Flows*,
Lect. Notes Phys. **681**, 131–132 (2005)
www.springerlink.com

For $\nu_0 = 1/2$, or $\sigma_a \approx 1$, the profile of a is thus a straight line, which agrees well with the hypotheses of the "0D" projection presented in Sect. 4.2 and of a parabolic profile of ν_t. For other values, the slope is no longer uniform, but we can estimate it here by its value at the center, a'_0. The integration constant a'_0 is determined by the conditions in the limits on a: if we normalize the total variation of a over the mixing zone thickness to 1, then:

$$a'_0 = \left(\int_{-1}^{+1} \left[1 - \xi^2\right]^{\frac{1}{2\nu_0} - 1} d\xi \right)^{-1} = \frac{\Gamma[1/(2\nu_0) + 1/2]}{\sqrt{\pi}\,\Gamma[1/(2\nu_0)]} \,, \tag{C.6}$$

where Γ here is the Euler Gamma function ($\Gamma[n+1] = n!$, when n is integer).

As was discussed in Sect. 5.2, the width of the mixing zone in the KH case must be defined based on the profile of the velocity U_y (not the profile of $\alpha^\pm$ or $C^\pm$), and thus we must substitute 1 for σ_c in the corresponding expressions of Table 5.1, p. 52 and in K_I in particular. According to (C.6), the length L_c of the concentration gradient profile at the center of the zone is then expressed as:

$$\frac{L_c}{L} \approx \frac{\sqrt{\pi}\,\Gamma[\sigma_c]}{2\,\Gamma[\sigma_c + 1/2]} \approx 1.25 \text{ for } \sigma_c = .7 \,. \tag{C.7}$$

D

Turbulent Flux of k Expressed in Two-Fluid Terms

Examining the general expression for the turbulent flux of a per mass quantity a in (7.5), we see that in order to express the flux $\overline{\varrho k u_i''}$, we must estimate the quantities $\overline{c^\pm \varrho k}$. We can reduce these to the turbulent kinetic energies in the coordinate system of each fluid, $\overline{c^\pm \varrho k^\pm}$, as defined by (3.10):

$$
\begin{aligned}
c^\pm \varrho k &= c^\pm \varrho \left(\boldsymbol{u}^\pm + \boldsymbol{U}^\pm - \boldsymbol{U} \right)^2 / 2 \\
&= c^\pm \varrho \left(\boldsymbol{u}^\pm \pm C^\mp \delta \boldsymbol{U} \right)^2 / 2 \\
&= c^\pm \varrho\, k^\pm \pm c^\pm \varrho\, \boldsymbol{u}^\pm C^\mp \delta \boldsymbol{U} + c^\pm \varrho\, (C^\mp \delta \boldsymbol{U})^2 / 2 \,,
\end{aligned}
\tag{D.1}
$$

where we used relation (3.8). We thus obtain after ensemble averaging:

$$
\frac{\overline{c^\pm \varrho k}}{\overline{c^\pm \varrho}} = \frac{\overline{c^\pm \varrho k^\pm}}{\overline{c^\pm \varrho}} + (C^\mp \delta \boldsymbol{U})^2 / 2 \,,
\tag{D.2}
$$

and after multiplying by $\boldsymbol{u}^\pm$ and ensemble averaging:

$$
\overline{c^\pm \varrho k \boldsymbol{u}^\pm} = \overline{c^\pm \varrho k^\pm \boldsymbol{u}^\pm} \pm 2\, \frac{\overline{c^\pm \varrho k^\pm}}{\overline{c^\pm \varrho}}\, C^+ C^- \,\overline{\varrho}\, \delta \boldsymbol{U} \,.
\tag{D.3}
$$

These two expressions substituted into (7.5) with $a = k$ then lead to (7.7).

A. Llor: *Statistical Hydrodynamic Models for Developed Mixing Instability Flows*,
Lect. Notes Phys. **681**, 133–133 (2005)
www.springerlink.com

E

About the Effects of Compressibility on Closures of Turbulent Fluxes

Boussinesq–Reynolds type closures are appropriate for quantities which, in a first approximation, are *transported by the material* and have *gradient lengths larger than the internal scale of turbulence* [61,77, §4.7]. For an incompressible fluid, this would apply to the mass fraction of a pollutant, the momentum, and the internal energy. Thus the turbulent coefficients of diffusion, viscosity, and thermal conduction can be introduced, all of which are homogeneous to $[L]^2[T]^{-1}$ and related by Schmidt–Prandtl numbers. However, these closures must be adapted for DAM applications, which involve the effects of compressibility and shocks (in the latter case, limiters are also necessary) [5,6,139,140]. We shall now examine the example of internal energy.

In a compressible fluid, the internal energy is not a good passive scalar, because there are exchanges through pressure work. We must then resort to the *entropy*, which is affected only by viscosity and thermal diffusion (which are disregarded here, but could be incorporated). If $e = \mathcal{E}(v, s)$ is the equation of state of the fluid, we define the average values $\hat{e}$, $\hat{p}$, and $\hat{T}$ and the associated fluctuations using the relations:

$$\hat{e} = \mathcal{E}(\tilde{v}, \tilde{s}) , \tag{E.1a}$$

$$-\hat{p} = \left.\frac{\partial \mathcal{E}}{\partial v}\right|_s (\tilde{v}, \tilde{s}) , \tag{E.1b}$$

$$\hat{T} = \left.\frac{\partial \mathcal{E}}{\partial s}\right|_v (\tilde{v}, \tilde{s}) \tag{E.1c}$$

$$a^{\iota} = a - \hat{a} . \tag{E.1d}$$

We then express the turbulent flux of energy as a function of the turbulent fluxes of mass and entropy in order to apply the Boussinesq–Reynolds closure to the latter:

$$\overline{\varrho e u_i''} = \overline{\varrho e^{\iota} u_i''} = \hat{T}\,\overline{\varrho s'' u_i''} - \hat{p}\,\overline{\varrho v'' u_i''} + \mathcal{O}(2)$$
$$\overset{\mathrm{m}}{=} -\hat{T}\,\overline{\varrho}\,D_t^s\,(\tilde{s})_{,i} - \hat{p}\,\overline{u_i''} = -\overline{\varrho}\,D_t^s\,((\hat{e})_{,i} + \hat{p}\,(\tilde{v})_{,i}) - \hat{p}\,\overline{u_i''} , \tag{E.2}$$

A. Llor: *Statistical Hydrodynamic Models for Developed Mixing Instability Flows*,
Lect. Notes Phys. **681**, 135–136 (2005)
www.springerlink.com

where, a priori, $D_t^s \approx D_t^c$. With the approximations $\hat{e} \approx \tilde{e}$ and $\hat{p} \approx \overline{p}$ valid in second order, we obtain:

$$\overline{\varrho e u_i''} \stackrel{\text{m}}{=} -\overline{\varrho}\, D_t^s\, (\tilde{e})_{,i} + \overline{p}\left(D_t^s\, \frac{(\overline{\varrho})_{,i}}{\overline{\varrho}} - \overline{u_i''}\right).$$

$$(\text{E.3})$$

Compared with the standard Boussinesq–Reynolds closure, there is now a corrective term, often called "adiabatic gradient correction," which is related to the work of the pressure forces during turbulent diffusion, and which introduces both the density gradient and the turbulent mass flux.[1] The modeling of this latter flux is discussed in Sect. 7.2.

The adiabatic gradient corrections do not seem to have been used for single-fluid compressible turbulence [60, 63], probably because they provide only minor corrections in aerodynamic flows. However, they are mandatory in the presence of shocks: indeed, a weak shock introduces a discontinuous, but quasi-isentropic, variation in the density, and the adiabatic correction thus entirely corrects the singularity of the fluxes. For consistency, the set of closures naturally requires adiabatic corrections for all the quantities modified by the volume variations, by adapting the pressure $\overline{p}$. For example, in the case of the *purely advective* part of the turbulent fluxes of $\tilde{k}$ and $\tilde{\varepsilon}$ (the pressure–velocity correlations $\overline{pu_i'}$ are excluded), the effective pressures to be applied in (E.3) are, respectively, $2/3\overline{\varrho}\tilde{k}$ and $(2C_{\varepsilon 1}/3 + C_{\varepsilon 3})\overline{\varrho}\tilde{\varepsilon}$.

In certain models of mixing flows, corrections of this type have already been introduced for the turbulent mass flux [13–17]. The effect of the adiabatic gradient on the "enthalpic production" term of turbulence is not really perceptible except for the production of $\tilde{k}$ due to shocks [142]: the behavior of the model seems to be more "suitable," but at the cost of reduced robustness. However, this term cannot resolve the inconsistency problems of the dissipative circuit in DAM's k–ε model as discussed in Sect. 7.2.

Incidentally, as already noticed [175, 176], there does not seem to be any justification for disregarding the internal energy flux associated with the "Rayleigh–Taylor production" term $\overline{pu_i''}$, as mentioned in item 12 of Sect. 2.3. However, if we apply the "adiabatic gradient" closure to the turbulent flux of internal energy, we obtain a term that compensates exactly for $\overline{pu_i''}$, as shown by formula (E.3).

---

[1] This correction for the work of pressure in the turbulent flux of energy appears to be intuitive in the model of an isentropic atmosphere, where there is a vertical energy gradient related to the pressure variation: despite the presence of turbulent motions that mix the gas, entropy remains uniform and thus there should not be a mean energy flux, which contradicts the classic Boussinesq–Reynolds closure (we disregard the dissipations and all the phenomena specific to meteorological flows).

References

1. *Chocs*, Revue Scientifique et Technique de la Direction des Applications Militaires, CEA, **14** (1995).
2. V.A. Andronov, S.M. Bakhrakh, V.N. Mokhov, V.V. Nikiforov, A.V. Pevnitskii, "Effect of turbulent mixing on the compression of laser targets", *Pis'ma Zh. Eksp. Teor. Fiz.* **1**,62 (1979) or *Sov. Phys. JETP Lett.* **29**,57 (1979).
3. M. Bonnet, S. Gauthier, P. Spitz, DAM Internal Report,[1] 3/31/88.
4. M. Bonnet, S. Gauthier, G. Samba, DAM Internal Report,[1] 4/6/90.
5. M. Bonnet, S. Gauthier, P. Spitz, "Numerical simulations with a 'k–ε' mixing model in the presence of shock waves", p.397 in: W.P. Dannevik, A.C. Buckingham, C.E. Leith, Advances in Compressible Turbulent Mixing, LLNL Conf-8810234, 1992 (Proceedings of the 1st International Workshop on the Physics of Compressible Turbulent Mixing, Princeton, New Jersey, 10/24/88).
6. M. Bonnet, S. Gauthier, "A k–ε model for turbulent mixing in shock-tube flows induced by Rayleigh–Taylor instability", *Phys. Fluids A* **2**,1685 (1990).
7. A. Llor, in: P. Tolla, DAM Internal Report,[1] 11/16/99.
8. S. Gauthier, DAM Internal Report,[1] 9/22/82.
9. S. Gauthier, B. Sitt, DAM Internal Report,[1] 1/17/83.
10. A. Froger, S. Gauthier, DAM Internal Report,[1] 9/24/85.
11. M. Bonnet, S. Gauthier, J.-F. Haas, P. Spitz, DAM Internal Report,[1] 8/6/87.
12. C.E. Leith, "Development of a two-equation turbulent mix model", LLNL, Report UCRL-96036, 12/86.
13. V.A. Andronov, S.M. Bakhrakh, E.E. Meshkov, V.V. Nikiforov, A.V. Pevnitskii, A.I. Tolshmyakov, "An experimental investigation and numerical modeling of turbulent mixing in one-dimensional flows", *Dokl. Akad. Nauk SSSR* **264**,76 (1982) or *Sov. Phys. Dokl.* **27**,393 (1983).

---

[1] © Although DAM internal reports are not accessible outside the DAM, most of their authors can be contacted by e-mail at the following addresses:
didier.besnard@cea.fr, daniel.bouche@cea.fr, catherine.cherfils@cea.fr,
alain.froger@cea.fr, serge.gauthier@cea.fr, olivier.gregoire@cea.fr,
jean-francois.haas@cea.fr, antoine.llor@cea.fr,
roland.omnes@th.u-psud.fr, bernard.rebourcet@cea.fr,
gerald.samba@cea.fr, bernard.sitt@cea.fr, denis.souffland@cea.fr,
patrick.spitz@cea.fr, eric.vanrenterghem@cea.fr.

14. V.V. Nikiforov, "Calculation of gravitational mixing in non-automodel flows", p.478 in: P.F. Linden, D.L. Youngs, S.B. Dalziel, Proceedings of the 4th International Workshop on the Physics of Compressible Turbulent Mixing, Cambridge, Royaume Uni, 3/29/93.

15. V.A. Andronov, I.G. Zhidov, E.E. Meshkov, N.V. Nevmerzhitskii, V.V. Nikiforov, A.N. Razin, V.G. Rogatchev, A.I. Tolshmyakov, Yu.V. Yanilkin, "Computational and experimental studies of hydrodynamic instabilities and turbulent mixing (review of NVIIEF efforts)", Los Alamos National Laboratory, LA-12896 (1995).

16. V.A. Andronov, S.A. Grushin, V.V. Nikiforov, A.N. Razin, O.M. Velichko, Yu.A. Yudin (editor D.C. Besnard), DAM Internal Report,[1] 7/4/95.

17. D. Souffland, DAM Internal Report,[1] 7/28/98.

18. A.V. Polyonov, "The heterogeneous k–ε model of gravitational mixing", in: V. Rupert, Collected papers presented at the international Workshop on Richtmyer–Meshkov and Rayleigh–Taylor Mixing, Pleasanton, Californie, 11/89.

19. A.V. Polyonov, "Shock-induced mixing, convergence and heat conduction within the framework of heterogeneous k–ε model", p.495 in: Proceedings of the 3rd International Workshop on the Physics of Compressible Turbulent Mixing, Royaumont, France, 6/17/91.

20. V.E. Neuvazhayev, A.V. Polyonov, V.G. Yakovlev, "Description of transitional layer effect in simulations with k–ε model", p.444 in: P.F. Linden, D.L. Youngs, S.B. Dalziel, Proceedings of the 4th International Workshop on the Physics of Compressible Turbulent Mixing, Cambridge, Royaume Uni, 3/29/93.

21. A.V. Polyonov, V.G. Yakovlev, "Mixture separation in the framework of Hk–ε model", p.497 in: P.F. Linden, D.L. Youngs, S.B. Dalziel, Proceedings of the 4th International Workshop on the Physics of Compressible Turbulent Mixing, Cambridge, Royaume Uni, 3/29/93.

22. D.C. Besnard, F.H. Harlow, "Turbulence in two-field incompressible flow", LANL, Report LA-10187-MS, 5/85.

23. D.C. Besnard, F.H. Harlow, R. Rauenzahn, "Conservation and transport properties of turbulence with large density variations", LANL, Report LA-10911-MS, 2/87.

24. D.C. Besnard, F.H. Harlow, "Turbulence in multiphase flow", *Int. J. Multiphase Flow* **14**,679 (1988).

25. D.C. Besnard, F.H. Harlow, R. Rauenzahn, C. Zemach, "Turbulence transport equations for variable-density turbulence and their relationship to two-field models", LANL, Report LA-12303-MS, 6/92.

26. O. Grégoire, "Modèle multiéchelle pour les écoulements turbulents et compressibles en présence de forts gradients de densité" [Multiscale model for turbulent and compressible flows in the presence of strong density gradients], Thèse de Doctorat, Université d'Aix–Marseille II, 2/11/97.

27. O. Grégoire, D. Souffland, DAM Internal Report,[1] 5/18/98.

28. O. Grégoire, DAM Internal Report,[1] 1/18/00.

29. M.J. Andrews, "Turbulent mixing by Rayleigh–Taylor instability", PhD thesis, Imperial College, Report CFDU 86/10 (1986).

30. M.J. Andrews, "An experimental study of turbulent mixing by Rayleigh–Taylor instabilities and a two-fluid model of the mixing phenomena", p.7 in: W.P. Dannavik, A.C. Buckingham, C.E. Leith, Advances in Compressible Turbulent

Mixing, LLNL Conf-8810234, 1992 (Proceedings of the 1st International Workshop on the Physics of Compressible Turbulent Mixing, Princeton, New Jersey, 10/24/88).

31. D.L. Youngs, "Numerical simulation of turbulent mixing by Rayleigh–Taylor instability", *Physica D*, **12**,32 (1984).

32. D.L. Youngs, "Modelling turbulent mixing by Rayleigh–Taylor instability", *Physica D*, **37**,270 (1989).

33. D.L. Youngs, "Numerical simulation of mixing by Rayleigh–Taylor and Richtmyer–Meshkov instabilities", *Laser and Particle Beams*, **12**,725 (1994).

34. C.W. Cranfill, "A multifluid turbulent-mix model", LANL, Report LA-UR-91-403, 1/30/91.

35. C.W. Cranfill, "A new multifluid turbulent-mix model", LANL, Report LA-UR-92-2484, 8/3/92.

36. Y. Chen, J. Glimm, D.H. Sharp, Q. Zhang, "A two-phase model of the Rayleigh–Taylor mixing zone", *Phys. Fluids* **8**,816 (1996).

37. J. Glimm, D. Saltz, D.H. Sharp, "A general closure relation for incompressible mixing layers induced by interface instabilities", p.179 in: G. Jourdan, L. Houas, Proceedings of the 6th international workshop on the physics of compressible turbulent mixing, Marseille, France, 6/18–6/21/97, IUSTI/CNRS, Marseille, 1997.

38. D. Saltz, W. Lee, T.-R. Hsiang, "Two-phase flow analysis of unstable fluid mixing in one-dimensional geometry", *Phys. Fluids* **12**,2461 (2000).

39. J. Glimm, D.H. Sharp, "Chaotic mixing as a renormalization-group fixed point", *Phys. Rev. Lett.* **64**,2137 (1990).

40. J. Glimm, Q. Zhang, D.H. Sharp, "The renormalization-group dynamics of chaotic mixing of unstable interfaces", *Phys. Fluids A* **3**,1333 (1991).

41. Q. Zhang, "Analytical solutions of Layzer-type approach to unstable interfacial fluid mixing", *Phys. Rev. Lett.* **81**,3391 (1998).

42. D. Ofer, D. Shvarts, Z. Zinamon, S.A. Orszag, "Mode coupling in nonlinear Rayleigh–Taylor instability", *Phys. Fluids B* **4**,3549 (1992).

43. U. Alon, D. Shvarts, D. Mukamel, "Scale-invariant regime in Rayleigh–Taylor bubble-front dynamics", *Phys. Rev. E* **48**,1008 (1993).

44. U. Alon, J. Hecht, D. Mukamel, D. Shvarts, "Scale invariant mixing rates of hydrodynamically unstable interfaces", *Phys. Rev. Lett.* **72**,2867 (1994).

45. J. Hecht, U. Alon, D. Shvarts, "Potential flow models of Rayleigh–Taylor and Richtmyer–Meshkov bubble fronts", *Phys. Fluids* **6**,4019 (1994).

46. U. Alon, J. Hecht, D. Ofer, D. Shvarts, "Power laws and similarity of Rayleigh–Taylor and Richtmyer–Meshkov mixing fronts at all density ratios", *Phys. Rev. Lett.* **74**,534 (1995).

47. D. Ofer, U. Alon, D. Shvarts, R.L. McCrory, C.P. Verdon, "Modal model for the nonlinear multimode Rayleigh–Taylor instability", *Phys. Plasmas* **3**,3073 (1996).

48. D. Shvarts, U. Alon, D. Ofer, C.P. Verdon, R.L. McCrory, "Nonlinear evolution of multimode Rayleigh–Taylor instability in two and three dimensions", *Phys. Plasmas* **2**,2465 (1995).

49. J.C.V. Hansom, P.A. Rosen, T.J. Goldack, K. Oades, P. Fieldhouse, N. Cowperthwaite, D.L. Youngs, N. Mawhinney, A.J. Baxter, "Radiation driven planar foil instability and mix experiments at the AWE HELEN laser", *Laser and Particle Beams*, **8**,51 (1990), Annexe.

50. J.D. Ramshaw, "Simple model for linear and nonlinear mixing at unstable fluid interfaces with variable acceleration", *Phys. Rev. E* **58**,5834 (1998).

51. G. Dimonte, "Spanwise homogeneous buoyancy–drag model for Rayleigh–Taylor mixing and experimental evaluation", *Phys. Plamas* **7**,2255 (2000).

52. A. Llor, in: P. Tolla, DAM Internal Report,[1] 6/7/00.

53. A. Vallet, O. Simonin, A. Llor, "k-ε-Σ: a two-fluid model of turbulent mixing flows induced by Kelvin–Helmholtz, Rayleigh–Taylor and Richtmyer–Meshkov instabilities", Symposium on Trends in Numerical and Physical Modeling for Industrial Multiphase Flows, Cargèse, France, 9/27/00.

54. A. Llor, "Response of turbulent RANS models to self-similar variable acceleration RT mixing: an analytical '0D' analysis", in: O. Schilling, Proceedings of the 8th international workshop on compressible turbulent mixing, Pasadena, California, 12/09–12/14/01, LLNL Report UCRL-MI-146350, 2002.

55. A. Llor, "Bulk turbulent transport and structure in Rayleigh–Taylor, Richtmyer–Meshkov and variable acceleration instabilities", *Laser and Particle Beams*, **21**,305 (2003).

56. O. Grégoire, DAM Internal Report,[1] 1/13/00.

57. D.L. Youngs, A. Llor, "Preliminary results of LES simulations of self-similar variable acceleration RT mixing flows", in: O. Schilling, Proceedings of the 8th international workshop on compressible turbulent mixing, Pasadena, California, 12/09–12/14/01, LLNL Report UCRL-MI-146350, 2002.

58. A. Favre, "Equations fondamentales des fluides à masse volumique variable en écoulements turbulents" [Fundamental equations of fluids with variable density in turbulent flows], in: A. Favre, L.S.G. Kovasznay, R. Dumas, J. Gaviglio, M. Coantic, "La turbulence en mécanique des fluides" [Turbulence in fluid mechanics], Gauthier–Villars, Paris, 1976.

59. H. Tennekes, J.L. Lumley, "A first course in turbulence", MIT Press, Cambridge, 1972.

60. S.K. Lele, "Compressibility effects on turbulence", *Annu. Rev. Fluid Mech.* **26**,211 (1994).

61. R. Schiestel, "Modélisation et simulation des écoulements turbulents" [Modeling and simulation of turbulent flows], Hermès, Paris, 1993.

62. B.E. Launder, "An introduction to single-point closure methodology", in: T.B. Gatski, M.Y. Hussaini, J.L. Lumley, "Simulation and modeling of turbulent flows", Oxford, New York, 1996.

63. C.G. Speziale, "Modeling of turbulent transport equations", in: T.B. Gatski, M.Y. Hussaini, J.L. Lumley, "Simulation and modeling of turbulent flows", Oxford, New York, 1996.

64. D. Souffland, DAM Internal Report,[1] 1/11/93.

65. D. Souffland, DAM Internal Report,[1] 12/29/95.

66. D.C. Besnard, DAM Internal Report,[1] 4/16/87.

67. D. Bouche, D.C. Besnard, DAM Internal Report,[1] 12/15/95.

68. D. Souffland, O. Grégoire, DAM Internal Report,[1] 1/11/99.

69. B.I. Davydov, "On statistical dynamics of an incompressible turbulent fluid", *Dokl. Akad. Nauk CCCP.* **136**,47 (1961) or *Sov. Phys. Dokl.* **6**,10 (1961).

70. F.H. Harlow, P.I. Nakayama, "Turbulence transport equations", *Phys. Fluids* **10**,2323 (1967).

71. F.H. Harlow, P.I. Nakayama, "Transport of turbulence energy decay rate", LANL, Report LA-3854 (1968).

72. B.J. Daly, F.H. Harlow, "Transport equations in turbulence", *Phys. Fluids* **13**,2634 (1970).

73. C.W. Hirt, "Generalized turbulence transport equations", International seminar of the international centre for heat and mass transfer, Herceg Novi, Yugoslavie, 9/69.

74. K.H. Ng, D.B. Spalding, "Turbulence model for boundary layers near walls", *Phys. Fluids* **15**,20 (1972).

75. W.P. Jones, B.E. Launder, "The prediction of laminarisation with a two-equations model of turbulence", *Int. J. Heat Mass Transfer* **15**,301 (1972).

76. K. Hanjalic, B.E. Launder, "A Reynolds stress model of turbulence and its application to thin shear flows", *J. Fluid Mech.* **52**,609 (1972).

77. J.L. Lumley, "Some comments on turbulence", *Phys. Fluids A* **4**,203 (1992).

78. J.L. Lumley, B. Khajeh-Nouri, "Computational modeling of turbulent transport", in: Proceedings of the 2nd IUGG-IUTAM Symposium on Atmospheric Diffusion in Environmental Pollution, *Adv. in Geophys.* **18A**,169 (1974).

79. W. Kollmann, D. Vandromme, *Int. J. Heat Mass Transfer* **22**,1557 (1979).

80. D. Vandromme, H. Ha-Minh, J.R. Viegas, M.W. Rubesin, W. Kollmann "Second order closure for the calculation of compressible wall-bounded flows with an implicit Navier-Stokes solver", IVth Int. Symposium on Turbulent Shear Flows, Karlsruhe (1983).

81. J. Von Neumann, R.D. Richtmyer, *J. Appl. Phys.*, **21**,232 (1950);
R.D. Richtmyer, K.W. Morton, "Difference methods for initial-value problems", 2nd ed., Wiley, New York, 1967 (chapitres 12.10 à 12.13);
D.J. Benson, "Computational methods in Lagrangian and Eulerian hydrocodes", *Comp. Meth. in Appl. Mech. and Eng.*, **99**,235 (1992).

82. M. Ishii, "Thermo-fluid dynamic theory of two-phase flow", Eyrolles, Paris, 1975.

83. V.H. Ransom, D.L. Hicks, "Hyperbolic two-pressure models for two-phase flow", *J. Comput. Phys.* **53**,124 (1984).

84. H.B. Stewart, B. Weindroff, "Two-phase flows: models and methods", *J. Comput. Phys.* **56**,363 (1984).

85. V.H. Ransom, "Numerical modelling of two-phase flows", Cours de l'Ecole d'Eté d'Analyse Numérique [Lecture at the Summer School on Numerical Analysis], CEA-INRIA-EDF, 6/12-6/23/89.

86. I. Toumi, A. Kumbaro, H. Paillere, "Approximate Riemann solvers and flux vector splitting schemes for two-phase flows", in: "30th computational fluid dynamics", Lecture Series 1999-03, Von Karman Institute for Fluid Dynamics, 8-3/12/99, §2.

87. J.C.R. Hunt, R.J. Perkins, J.C.H. Fung, "Review of the problems of modelling disperse two-phase flows", in: G.F. Hewitt, J.H. Kim, R.T. Lahey, J.M. Delhaye, N. Zuber, "Multiphase science and technology, Vol. 8", Begell, New York, 1994.

88. C. Morel, "Modélisation multidimensionnelle des écoulements diphasiques gaz-liquide. Application à la simulation des écoulements à bulles ascendants en conduite verticale" [Multidimensional modeling of two-phase gas-liquid flows. Application to the simulation of rising bubble flows in vertical ducts], Thèse de Doctorat, Ecole Centrale de Paris, 10/31/97 (and references cited therein).

89. O. Simonin, "Modélisation numérique des écoulements turbulents diphasiques à inclusions dispersées" [Numerical modeling of two-phase turbulent flows with dispersed inclusions], Cours à l'Ecole de Printemps CNRS de Mécanique des

Fluides Numérique [Lecture at the CNRS Spring School on Numerical Fluid Mechanics], Aussois, 1991.

90. O. Simonin, "Continuum modelling of dispersed two-phase flows", in: "Combustion and turbulence in two-phase flows", Lecture series 1996-02, Von Karman Institute for Fluid Dynamics, 1/29–2/2/96.

91. D.L. Youngs, personal communication, 7/4/00.

92. S. Chandrasekhar, "Hydrodynamic and hydromagnetic stability", Clarendon, Oxford, 1961 et Dover, New York, 1981.

93. D.H. Sharp, "An overview of Rayleigh–Taylor instability", *Physica* **12D**,3 (1984).

94. R. Omnès, DAM Internal Report,[1] 12/20/85.

95. H.J. Kull, "Theory of the Rayleigh–Taylor instability", *Phys. Rep.* **206**,197 (1991).

96. D. Bouche, S. Gauthier, D. Souffland, DAM Internal Report,[1] 6/10/97.

97. N.A. Inogamov, "The role of Rayleigh–Taylor and Richtmyer–Meshkov instabilities in astrophysics: an introduction", *Astrophys. Space Phys. Rev.* **10**,1 (1999).

98. K.O. Mikaelian, "Turbulent energy at accelerating and shocked interfaces", *Phys. Fluids A* **2**,592 (1990).

99. K.O. Mikaelian, "Kinetic energy of Rayleigh–Taylor and Richtmyer–Meshkov instabilities", *Phys. Fluids A* **3**,2625 (1991).

100. P.B. Spitz, J.-F. Haas, "Numerical calibration of Rayleigh–Taylor induced turbulent flows with a k-ε mix model", p.511 in: Proceedings of the 3rd International Workshop on the Physics of Compressible Turbulent Mixing, Royaumont, France, 6/17/91.

101. C. Cherfils, "Comparison of two turbulent diffusion dissipation models via similarity methods", p.413 in: P.F. Linden, D.L. Youngs, S.B. Dalziel, Proceedings of the 4th International Workshop on the Physics of Compressible Turbulent Mixing, Cambridge, Royaume Uni, 3/29/93.

102. C. Cherfils, A.K. Harrison, "Comparison of different statistical models of turbulence by similarity methods", UCRL-JC-117404, Lawrence Livermore National Laboratory, May 94, or Proceedings of the Summer Meeting, ASME Fluids Engineerieng Division, Incline, Nevada, 6/19/94.

103. C. Cherfils, DAM Internal Report,[1] 1/25/95.

104. G.I. Barenblatt, "Selfsimilar turbulence propagation from an instantaneous plane source", p.48 in: G.I. Barenblatt, G. Iooss, D.D. Joseph, "Nonlinear dynamics and turbulence", Pittman, Boston, 1983.

105. G.I. Barenblatt, "Scaling, self-similarity, and intermediate asymptotics", Cambridge, Boston, 1996, §10.2.4.

106. V.E. Neuvazhayev, V.G. Yakovlev, "Numerical calculation of interfaces turbulent mixing by Rayleigh–Taylor instability on the basis of semi-empirical models", in: V. Rupert, Collected papers presented at the international Workshop on Richtmyer–Meshkov and Rayleigh–Taylor Mixing, Pleasanton, Californie, 11/89.

107. V.E. Neuvazhayev, "Properties of the k-ε model for turbulent mixing", p.449 in: P.F. Linden, D.L. Youngs, S.B. Dalziel, Proceedings of the 4th International Workshop on the Physics of Compressible Turbulent Mixing, Cambridge, Royaume Uni, 3/29/93.

108. N. Freed, D. Ofer, D. Shvarts, S.A. Orszag, "Similarity theory of incompressible random Rayleigh–Taylor instability", p.61 in: W.P. Dannavik, A.C. Buckingham, C.E. Leith, Advances in Compressible Turbulent Mixing, LLNL Conf-8810234, 1992 (Proceedings of the 1st International Workshop on the Physics of Compressible Turbulent Mixing, Princeton, New Jersey, 10/24/88).

109. N. Freed, D. Ofer, D. Shvarts, S.A. Orszag, "Two-phase flow analysis of self-similar turbulent mixing by Rayleigh–Taylor instability", *Phys. Fluids A* **3**,912 (1991).

110. I. Wygnanski, H.E. Fiedler, "The two-dimensional mixing region", *J. Fluid Mech.* **41**,327 (1970).

111. R.P. Patel, "An experimental study of a plane mixing layer", *AIAA J.* **11**,67 (1973).

112. G.L. Brown, A. Roshko, "On density effects and large-scale structure in turbulent mixing layers", *J. Fluid Mech.* **64**,775 (1974).

113. R.G. Batt, "Turbulent mixing of passive and chemically reacting species in a low-speed shear layer", *J. Fluid Mech.* **82**,53 (1977).

114. F.K. Browand, B.O. Latigo, "Growth of the two-dimensional mixing layer from a turbulent and nonturbulent boundary layer", *Phys. Fluids* **22**,1011 (1979).

115. D. Oster, I. Wygnanski, "The forced mixing layer between parallel streams", *J. Fluid Mech.* **123**,91 (1982).

116. B.O. Latigo, "Coherent structure interactions in a two-stream plane turbulent mixing layer with impulsive acoustic excitation", *Phys. Fluids A* **1**,1701 (1989).

117. R.D. Mehta, R.V. Westphal, "Effect of velocity ratio on plane mixing layer development", p.3.2.1 in: Proceedings of the 7th Symposium on Turbulent Shear Flows, Stanford University, Stanford, 8/21/89.

118. K.I. Read, "Experimental investigation of turbulent mixing by Rayleigh–Taylor instability", *Physica D* **12**,45 (1984).

119. Yu. Kucherenko, L.I. Shibarshov, V.I. Chitaikin, S.I. Balabin, A.P. Pylaev, "Experimental study of the gravitational turbulent mixing self-similar mode", p.427 in: Proceedings of the 3rd International Workshop on the Physics of Compressible Turbulent Mixing, Royaumont, France, 6/17/91.

120. G. Dimonte, M. Schneider, "Turbulent Rayleigh–Taylor instability experiments with variable acceleration", *Phys. Rev. E* **54**,3740 (1996).

121. G. Dimonte, M. Schneider, "Density ratio dependence of Rayleigh–Taylor mixing for sustained and impulsive acceleration histories", *Phys. Fluids* **12**,304 (2000).

122. S.B. Daziel, P.F. Linden, D.L. Youngs, "Self-similarity and internal structure of turbulence induced by Rayleigh–Taylor instability", *J. Fluid Mech.*, **399**,1 (1999).

123. G. Dimonte, D.L. Youngs, A. Dimits, S. Weber, M. Marinak, S. Wunsch, C. Garasi, A. Robinson, M.J. Andrews, P. Ramaprabhu, A.C. Calder, B. Fryxell, J. Biello, L. Dursi, P. MacNeice, K. Olson, P. Ricker, R. Rosner, F. Timmes, H. Tufo, Y.-N. Young, M. Zingale, "A comparative stydy of the turbulent Rayleigh–Taylor instability using high-resolution three-dimensional numerical simulations: the Alpha-Group collaboration", *Phys. Fluids* **16**,1668 (2004).

124. G. Dimonte, M. Schneider, "Turbulent Richtmyer–Meshkov instability experiments with strong radiatively driven shocks", *Phys. Plasmas* **4**,4347 (1997) (and references cited therein for the exponent of self-similar growth of the Richtmyer–Meshkov instability).

144   References

125. K.-Y. Chien, R.E. Fergusson, A.L. Kuhl, H.M. Glaz, P. Colella, "Inviscid simulations of turbulent shear layers – mean flow analysis", p.161 in: W.P. Dannavik, A.C. Buckingham, C.E. Leith, Advances in Compressible Turbulent Mixing, LLNL Conf-8810234, 1992 (Proceedings of the 1st International Workshop on the Physics of Compressible Turbulent Mixing, Princeton, New Jersey, 10/24/88).

126. K.-Y. Chien, R.E. Fergusson, A.L. Kuhl, H.M. Glaz, P. Colella, "Inviscid simulations of turbulent shear layers – fluctuating flow profiles", p.4.3.1 in: Proceedings of the 7th Symposium on Turbulent Shear Flows, Stanford University, Stanford, 8/21/89.

127. A.A. Stadnik, V.P. Statsenko, Yu.V. Yanilkin, V.A. Zhmailo, "Direct numerical simulation of turbulent mixing in shear flows", p.381 in: R. Young, J. Glimm, B. Boston, Proceedings of the 5th international workshop on compressible turbulent mixing, Stony Brook, New York, 7/18–7/21/95, World Scientific, Singapore, 1995.

128. X.L. Li, "Study of three-dimensional Rayleigh–Taylor instability in compressible fluids through level set method and parallel computation", *Phys. Fluids A* **5**,1904 (1993).

129. A.A. Stadnik, V.P. Statsenko, Yu.V. Yanilkin, V.A. Zhmailo, "Direct numerical simulation of gravitational turbulent mixing", p.392 in: R. Young, J. Glimm, B. Boston, Proceedings of the 5th international workshop on compressible turbulent mixing, Stony Brook, New York, 7/18–7/21/95, World Scientific, Singapore, 1995.

130. O.G. Sin'kova, A.L. Stadnik, V.P. Statsenko, Yu.V. Yanilkin, V.A. Zhmailo, "Three-dimensional numerical simulation of gravitational turbulent mixing", p.470 in: G. Jourdan, L. Houas, Proceedings of the 6th international workshop on the physics of compressible turbulent mixing, Marseille, France, 6/18–6/21/97, IUSTI/CNRS, Marseille, 1997.

131. G. Jourdan, L. Houas, J.-F. Haas, G. Ben–Dor, "Thickness and volume measurements of a Richtmyer–Meshkov instability-induced mixing zone in a square shock tube", *J. Fluid Mech.* **349**,67 (1997) (and references cited therein).

132. J.K. Prasad, A. Rasheed, S. Kumar, B. Sturtevant, "The late-time development of the Richtmyer–Meshkov instability", *Phys. Fluids* **12**,2108 (2000) (et références citées).

133. A. Cook, B. Cabot, P. Miller, "The mixing transition in Rayleigh–Taylor instability", in: proceedings to be published of the 9th international workshop on compressible turbulent mixing, Cambridge, UK, 07/19–07/23/04.

134. H. Schlichting, K. Gersten, "Boundary layer theory", 8th éd., Springer, Berlin, 2000.

135. F. Poggi, "Analyse par vélocimétrie d'un mélange gazeux crée par instabilité de Richtmyer–Meshkov" [Velocimetric analysis of a gaseous mixture created by a Richtmyer–Meshkov instability], Thèse de Doctorat, Université de Poitiers, 12/1/97.

136. F. Poggi, M.H. Thorembey, G. Rodriguez, J.-F. Haas, "Velocity measurements in turbulent gaseous mixtures induced by Richtmyer–Meshkov instability", p.416 in: G. Jourdan, L. Houas, Proceedings of the 6th international workshop on the physics of compressible turbulent mixing, Marseille, France, 6/18–6/21/97, IUSTI/CNRS, Marseille, 1997.

137. D. Souffland, O. Grégoire, S. Gauthier, F. Poggi, J.M. Kœnig, "Measurements and simulation of the turbulent energy levels in mixing zones generated in shock

tubes", p.486 in: G. Jourdan, L. Houas, Proceedings of the 6th international workshop on the physics of compressible turbulent mixing, Marseille, France, 6/18–6/21/97, IUSTI/CNRS, Marseille, 1997.

138. F. Poggi, M.H. Thorembey, G. Rodriguez, "Velocity measurements in turbulent gaseous mixtures induced by Richtmyer–Meshkov instability", *Phys. Fluids* **10**,2698 (1998).

139. E. Van Renterghem, DAM Internal Report,[1] 3/5/99.

140. E. Van Renterghem, DAM Internal Report,[1] 3/10/99.

141. A.V. Wilchinsky, K. Hutter, "On thermodynamic consistency of turbulent closures", *Thoer. Comput. Fluid Dynamics*, **15**,23 (2001).

142. D. Souffland, DAM Internal Report,[1] 3/31/00.

143. J.R. Ristorcelli, "A representation for the turbulent mass flux contribution to Reynolds stress and two-equation closures for compressible turbulence", NASA Langley Research Center, Report ICASE 93-88 (1993).

144. O. Zeman "A new model for super/hypersonic turbulent boundary layers", AIAA, Report 93-0897 (1993).

145. O. Grégoire, DAM Internal Report,[1] 12/1/00.

146. W. Rodi, "A new algebraic relation for calculating the Reynolds stresses", *ZAMM* **56**,219 (1976).

147. J. Gaviglio, "La turbulence dans les écoulements compressibles des gaz" [Turbulence in compressible gas flows], in: A. Favre, L.S.G. Kovasznay, R. Dumas, J. Gaviglio, M. Coantic, "La turbulence en mécanique des fluides" [Turbulence in fluid mechanics], Gauthier–Villars, Paris, 1976.

148. O. Zeman, "Dilatation dissipation: the concept and application in modelling compressible mixing layers", *Phys. Fluids A* **2**,178 (1990).

149. O. Zeman, "On the decay of compressible isotropic turbulence", *Phys. Fluids A* **3**,951 (1991).

150. H.S. Ribner, "Convection of a pattern of vorticity through a shock wave", NACA-R-1164 (1954);
H.S. Ribner, "Shock–turbulence interaction and the generation of noise", NACA-R-1233 (1955).

151. L. Jacquin, C. Cambon, E. Blin, "Turbulence amplification by a shock wave and rapid distortion theory", *Phys. Fluids A* **5**,2539 (1993).

152. D. Rotman, "Shock wave effects on a turbulent flow", *Phys. Fluids A* **3**,1792 (1991).

153. S. Lee, S.K. Lele, P. Moin, "Direct numerical simulation and analysis of shock turbulence interaction", Document AIAA, AIAA-91-0523 (1991).

154. S. Lee, P. Moin, S.K. Lele, "Interaction of isotropic turbulence with a shock wave", Department of Mechanical Engineering, Stanford University, California, Report TF-52 (1992).

155. S. Lee, "Effect of shock strength on shock turbulence interaction", Annual Research Briefs, Center for Turbulence Research, 329 (1993).

156. S. Lee, S.K. Lele, P. Moin, "Direct numerical simulations of isotropic turbulence interacting with a weak shock wave: effet of shock strength", *J. Fluid Mech.* **251**,533 (1993).

157. S. Lee, S.K. Lele, P. Moin, "Interaction of isotropic turbulence with shock waves: effet of shock strength", *J. Fluid Mech.* **340**,225 (1997).

158. B. Sitt, DAM Internal Report,[1] 5/31/89.

159. B. Rebourcet, "Une décomposition des discontinuités d'un système hyper-bolique multi-énergies dans le cadre des schémas décales en espace – appli-cation à la turbulence" [A decomposition of the discontinuities of a hyperbolic, multienergy system in the framework of space-shifted schemes: application to turbulence], Working Document, Février 1998.

160. Ya.B. Zel'dovich, Yu.P. Raizer, "Physics of shock waves and high-temperature hydrodynamic phenomena", Academic Press, New York, 1967 (Chaps. I and VII, and references cited therein).

161. D. Mihalas, B.W. Mihalas, "Foundations of radiation hydrodynamics", Oxford, New York, 1984 (Sect. 5.3).

162. A. Llor, DAM Internal Report,[1] 10/27/98 (and references cited therein).

163. M. Boucker, "Modélisation numérique multidimensionnelle d'écoulements diphasiques liquide–gaz en régimes transitoire et permanent: méthodes et ap-plications" [Multidimensional numerical modeling of two-phase liquid–gas flows in transient and steady regimes: methods and applications], Thèse de Doctorat, Ecole Normale Supérieure de Cachan, 12/3/98.

164. A. Llor, P. Bailly, "A new turbulent two-fluid RANS model for KH, RT and RM mixing layers", in: O. Schilling, Proceedings of the 8th international work-shop on compressible turbulent mixing, Pasadena, California, 12/09–12/14/01, LLNL Report UCRL-MI-146350, 2002.

165. A. Llor, P. Bailly, "A new turbulent two-field concept for modeling Rayleigh–Taylor, Richtmyer–Meshkov and Kelvin–Helmholtz mixing layers", *Laser and Particle Beams*, **21**,311 (2003).

166. A. Llor, P. Bailly, O. Poujade, "Derivation of a minimal 2–fluid 2–structure and 2–turbulence (2SFK) model for gravitationally induced turbulent mixing layers", in: proceedings to be published of the 9th international workshop on compressible turbulent mixing, Cambridge, UK, 07/19–07/23/04.

167. S. Gauthier, DAM Internal Report,[1] 2/26/98.

168. D.L. Youngs, "Representation of the molecular mixing process in a two-phase flow turbulent mixing model", p.321 in: R. Young, J. Glimm, B. Boston, Pro-ceedings of the 5th international workshop on compressible turbulent mixing, Stony Brook, New York, 7/18–7/21/95, World Scientific, Singapore, 1995.

169. P.F. Linden, J.M. Redondo, "Molecular mixing in Rayleigh–Taylor instability. Part I: Global mixing", *Phys. Fluids A*, **3**,1269 (1991).

170. P.F. Linden, J.M. Redondo, D.L. Youngs, "Molecular mixing in Rayleigh–Taylor instability", *J. Fluid Mech.*, **265**,97 (1994).

171. A. Llor, P. Bailly, O. Poujade, "Volume fraction profiles of transport struc-tures in Rayleigh-Taylor turbulent mixing zone: evidence of enhanced diffusion processes", in: proceedings to be published of the 9th international workshop on compressible turbulent mixing, Cambridge, UK, 07/19–07/23/04.

172. V.E. Neuvazhayev, "Properties of a model for the turbulent mixing of the boundary between accelerated liquids differing in density", *J. Appl. Mech. Tech. Phys.* **24**,680 (1983).

173. J.R. Ristorcelli, "A pseudo-sound constitutive relationship for the dilatational covariances in compressible turbulence", *J. Fluid Mech.* **347**,37 (1997).

174. W.T. Chu, L.S.G. Kovasznay, "Non-linear interactions in a viscous heat con-ducting compressible gaz", *J. Fluid Mechanics*, **3**,494 (1958).

175. B. Rebourcet, DAM Internal Report,[1] 1/13/94.

176. B. Rebourcet, DAM Internal Report,[1] 6/11/97.

Index

Lecture Notes in Physics

For information about earlier volumes
please contact your bookseller or Springer
LNP Online archive: springerlink.com